Studies in Computational Intelligence 430

Editor-in-Chief

Prof. Janusz Kacprzyk
Systems Research Institute
Polish Academy of Sciences
ul. Newelska 6
01-447 Warsaw
Poland
E-mail: kacprzyk@ibspan.waw.pl

For further volumes:
http://www.springer.com/series/7092

Roger Lee (Ed.)

Software Engineering Research, Management and Applications 2012

Editor
Roger Lee
Computer Science Department
Software Engineering & Information
Technology Institute
Central Michigan University
MI
USA

ISSN 1860-949X e-ISSN 1860-9503
ISBN 978-3-642-44119-6 ISBN 978-3-642-30460-6 (eBook)
DOI 10.1007/978-3-642-30460-6
Springer Heidelberg New York Dordrecht London

Printed on acid-free paper

Springer is part of Springer Science+Business Media (www.springer.com)

Preface

The purpose of the 10th ACIS International Conference on Software Engineering Research, Management and Applications (SERA 2012) held on May 30-June 1, 2012 in Shanghai, China was to bring together researchers and scientist, engineers, computer users, students to share their experiences and exchange new ideas, and research results about all aspects (theory, applications and tools) of software engineering, and to discuss the practical challenges encountered along the way and the solutions adopted to solve them. The conference organizers selected the best 14 papers from those papers accepted for presentation at the conference in order to publish them in this volume. The papers were chosen based on review scores submitted by members of the program committee, and underwent further rigorous rounds of review.

In Chapter 1, Economic globalization and Information Technology development have intensified the competition among modern enterprises, independent of the trade and sector in which they develop. In this intense battle, the customer relationship has arisen as an important resource for establishing a competitive position against the rest of competitors. Customer Relationship Management (CRM), particularly, is a new technology, strategy and idea that provides overall guidelines and support for business in the construction, maintenance and management of customer relationships. However, CRM implementation into Mexican Small and Medium Enterprises (SMEs), is, until now, a field under continuous exploration which mainly generates recommendations about success and failure factors. However, there is evidence about the success of CRM in developing countries that enables countries to establish a comparative basis. This paper provides an exploratory study about CRM experiences in developing countries' SMEs, and presents a framework which implements the minimal requirements identified for this kind of enterprises in Mexico.

In Chapter 2, View the explosion of data volume and high circulating on the web (satellite data, genomic data ...) the classification of the data (data mining technique) is required. The clustering was performed by a method based bio (social spiders) because there is currently no method of learning that can almost directly represent unstructured data (text). Thus, to make a good data classification must be a good representation of the data. The representation of these data is performed by a vector whose components are derived from the overall weight of the corpus used (TF-IDF). A language-independent method was used to represent text documents is that of n-grams characters and words. Several similarity measures have been tested. To validate the classification we used a measure of assessment based on recall and precision (f-measure).

In Chapter 3, Over the years, software has become ubiquitous in business processes from payroll systems to airline reservation systems. Software plays a vital role in facilitating business processes. Given the importance of these software systems, managing their quality is essential to the success of the business processes they support. Because quality attributes are important predictors of software quality as to provide a better understanding of the related features acquired for each software quality attribute, those features can be manipulated to improve the quality of a software project and determine the desired functional requirements necessary to satisfy the associated business processes. Accordingly, customer needs should be completely elicited in developing the software application then traced and referred back to throughout the software development process during the early requirements analysis phase. The proposed questionnaire empowers software developers to capture the functional reliability requirements and specify reliability related features for a software system.

In Chapter 4, Design patterns are good design solutions to recurring problems.

Many works were interested in design patterns identification either for reverse engineering purposes, or for design improvement purposes. All existing approaches considered that a pattern is detected through both its structure and behavior, but no one considers the semantic aspect conveyed by the class and method names. In this paper, we propose a technique that exploits the semantic aspect to identify occurrences of a pattern in a design. In addition to the structural and behavioral analyses, the semantic analysis is very useful, specifically when there is a doubt between two design patterns having similar structures. By resolving a non deterministic identification, the identification precision increases.

In Chapter 5, Organizational knowledge contributes to the requirements necessary for effective Enterprise Architecture (EA) design. The effectiveness of EA processes depend on the quality of both functional and non-functional requirements elicited during the EA design process. Existing EA frameworks consider EA design solely from a techno-centric perspective focusing on the interaction of business goals, strategies, and technology. However, many enterprises fail to achieve the business goals established for the EA because of miscommunication of stakeholder requirements. Though modeling functional and non-functional design requirements from a technical perspective better ensures delivery of EA, a more complete approach would take into account human behavior as a vital factor in EA design. The contribution of this paper is an EA design guideline based on human behavior and socio-communicative aspects of stakeholders and the enterprise using socio-oriented approaches to EA design and modeling.

In Chapter 6, Managing reconfigurable Distributed Real-time Embedded (DRE) systems is a tedious task due to the substantially increasing complexity of these systems and the difficulty to preserve their real-time aspect. In order to resolve this increasing complexity, we propose to develop a new middleware, called RCES4RTES (Reconfigurable Computing Execution Support for Real-Time Embedded Systems), allowing the dynamic reconfiguration of component-based DRE systems. This middleware provides a set of functions ensuring dynamic reconfiguration as well as monitoring and coherence of such systems using a small memory footprint and respecting real-time constraints.

In Chapter 7, Feature Modeling (FM) is an essential activity for capturing commonality and variability in software product lines. Most of today's FM tools are graphical and represent feature models as feature diagrams (FDs). Though FDs are intuitive at first sight, they generally lack of expressiveness and vague in syntax. To overcome these problems, some textual languages are proposed with a richer expressiveness and formal semantics. But are these languages superior than existing modeling approach as they stated, e.g., XML-based one, which is standard based and has vast of acceptance of application other than FM. In this paper, we elabo-rate the XML-based textual feature modeling approach, evaluate it from multi perspectives, and compare it with another representative textual FM language —TVL, a recently published language. We demonstrate the advantages and disadvantages of the XML-based approach, and argue that the XML-based approach is still the most available and pragmatic approach for adoption in industry though it has some clear limitations which can be fed to more advanced tool support.

In Chapter 8, The utility of a smart-phone application depends not only on its functionality but also on its key non-functional requirements (NFRs), such as ubiquity, safety and usability. Omissions or commissions of considerations of such NFRs could lead to undesirable consequences, such as lack of user safety and difficulty in using smart-phone features. Currently, however, there is little systematic methodology for dealing with NFRs for a smart-phone application, in consideration of the particular characteristics of smart-phones, such as limited screen-size and battery-life, and the availability of a variety of embedded sensors and input/output devices. In this paper, we propose a goal-oriented approach in which NFRs are treated as softgoals, and then used in exploring, and selecting among, alternative means for satisficing them. In this approach, both synergistic and antagonistic interactions among the softgoals are identified and analyzed, considering the particular characteristics of smart-phones. As a proof of concept, a fall detection and response feature of a smart-phone application is presented, along with a safety scenario.

In Chapter 9, Distributed objects usually refer to software modules that are designed to work together, but reside either in multiple computers connected via a network or in different processes inside the same computer. One object sends a message to another object in a remote machine or process to perform some task. Distributed objects are a potentially powerful tool that has only become broadly available for developers at large in the past few years. The power of distributing objects is not in the fact that a bunch of objects are scattered across the network. This paper presents an environment for supporting distributed application using shared object in Java within a heterogeneous environment for mobile applications. In particular, it offers a set of objects such as Lists, Queues, Stacks that can be shared across a network of heterogeneous machine in the same way as DSM systems. Shared is achieved without recourse to Java RMI or object proxies as in other object systems. An implementation of the environment MBO(Mobile Business Objects) is provided together with performance timings.

In Chapter 10, In recent years, with the growth of software engineering, agile software development methodologies have also grown substantially, replacing plan-driven approaches in many areas. Although prominent agile methodologies are in wide use today, there is no method which is suitable for all situations. It has therefore become essential to apply Situational Method Engineering (SME) approaches to produce agile methodologies that are tailored to fit specific software development situations. Since SME is a complex process, and there is a vast pool of techniques, practices, activities, and processes available for composing agile methodologies, tool support–in the form of Computer Aided Method Engineering (CAME) environments has become essential. Despite the importance of tool support for developing agile methodologies, available CAME environments do not fully support all the steps of method construction, and the need remains for a comprehensive environment. The Eclipse Process Framework Composer (EPFC) is an open-source situational method engineering tool platform, which provides an extensible platform for assembly-based method engineering in Eclipse. EPFC is fully extensible through provision of facilities for adding new method plug-ins, method packages, and libraries. The authors propose a plug-in for EPFC which enables method engineers to construct agile methodologies through an assembly-based SME approach. The plug-in provides facilities for the specification of the characteristics of a given project, selection of suitable agile process components from the method repository, and the final assembly of the selected method chunks, while providing a set of guidelines throughout the assembly process.

In Chapter 11, Business Intelligence (BI) is a set of tools, technologies and process in order to transform data into information and information to required knowledge for improve decision making in organization. Nowadays, we can confidently claim that the use of business intelligence solutions can increase the competitiveness of organization and outstanding it from other organization. This solution enables organization to use available information to exploit the competitive advantages of being a leader and have a better understanding of customer needs and demands to allow better communication with them. In this paper we explain about principals and elements of BI in fist section and in second section we discuss about the application of BI in banking industry and consider Saman Bank of Iran as a case study in order to applying BI solution.

In Chapter 12, Component-based development traditionally recognizes two parallel views (system-level view and component-level view), which correspond to two major concerns – development of a an application and development of a reusable component for the use in application development. By having different objectives, these views have relatively disparate notion of a component, which consequently means that they are difficult (yet necessary) to combine. In this paper, we propose a method (named CoDIT), which spans the gap between the two views by providing a synchronisation between system-level view (expressed in UML 2) and componentlevel view. For component-level view, the method supports component frameworks with method-call as the communication style. The variability in the composition mechanisms of the component frameworks is addressed by using principles of metacomponent systems. The benefits of the proposed method are evaluated on a real-life case study (in SOFA 2) together with measurements of development efforts.

In Chapter 13, Cloud resource architecture is put forward based on servers according to the characteristics of Cloud Computing. And then the trust degree for the resource scheduling in Cloud Computing is defined and a resource scheduling algorithm based on trust degree is presented. Finally, the functional characteristics of this scheduling algorithm are analyzed by simulation, and the simulation results show that the resource scheduling algorithm based on trust degree in Cloud computing possesses the better stability and low risk on completing tasks.

In Chapter 14, We present a location recognition system for anentertainment robot and propose an idea to design a sensor network space. The sensor space consists of CDS (Cadmium sulfide) sensor cell of 24 by 24. Also our implemented hardware system is tested for setting a reference value. It is important to have exact location recognition. The algorithm for acquiring the reference value is proposed and its performance is evaluated with data of the implemented hardware system.

It is our sincere hope that this volume provides stimulation and inspiration, and that it will be used as a foundation for works yet to come.

May 2012 Roger Lee

Contents

List of Contributors

Abdelmalek Amine
Dr Moulay Tahar University
of SAIDA, Algeria
E-mail: amine_abd1@yahoo.fr

Anahita Alipour
Sharif University of Technology, Iran
E-mail: alipour@ce.sharif.edu

Anthony White
Middlesex University, U.K.
E-mail: a.white@mdx.ac.uk

Bechir Zalila
University of Sfax, Tunisia
E-mail: bechir.zalila@enis.rnu.tn

Bing Wan
Hainan University, China
E-mail: wanbingglass@163.com

Brahim Hamid
University of Toulouse, France
E-mail: brahim.hamid@irit.fr

Chul-Ung Kang
JEJU National University, Korea
E-mail: cukang@jejunu.ac.kr

Dazhe Zhao
Neusoft Corporation, China
E-mail: zhaodz@neusoft.com

Dominic M. Mezzanotte Sr.
Towson University, USA
E-mail: dmezzanotte@towson.edu

Elli Georgiadou
Arab Academy for Science
& Technology, Egypt
E-mail: e.georgiadou@mdx.ac.uk

Fatma Krichen
University of Toulouse, France
E-mail: fatma.krichen@irit.fr

Garcia I
Technological University
of the Mixtec Region, Mexico
E-mail: ivan@mixteco.utm.mx

Haeng-Kon Kim
Catholic University of Daegu, Korea
E-mail: hangkon@cu.ac.kr

Hanêne Ben-Abdallah
University of Sfax, Tunisia
E-mail:
Hanene.BenAbdallah@fsegs.rnu.tn

Hongyuan Wang
Jilin University, USA
E-mail: hongyuan@jlu.edu.cn

Imène Issaoui
University of Sfax, Tunisia
E-mail: imene.issaoui@gmail.com

Jingang Zhou
Northeastern University, China
E-mail: zhou-jg@computer.org

Jiren Liu
Neusoft Corporation, China
E-mail: liujr@neusoft.com

Ji-Uoo Tak
Catholic University of Daegu, Korea
E-mail: lebbenle@cu.ac.kr

Jong-Hwan Lim
JEJU National University, Korea

Josh Dehlinger
Towson University, USA
E-mail: jdehlinger@towson.edu

Lawrence Chung
The University of Texas
at Dallas, USA
E-mail: chung@utdallas.edu

Li Xu
Neusoft Corporation, China
E-mail: xuli@neusoft.com

Lukas Hermann
Michal Malohlava Charles University,
Czech Republic
E-mail: hermann@d3s.mff.cuni.cz

Martinez A
RagaSoft Inc., Mexico
E-mail: alberto@ragasoft.com.mx

Marwa Abd Elghany
Arab Academy for Science
& Technology, Egypt
E-mail: marwam@aast.edu

Maryam Marefati
IAU University Arak Branch, Iran
E-mail: mmarefati@gmail.com

Mengxing Huang
Hainan University, China
E-mail: huangmx09@gmail.com

Michal Malohlava
Michal Malohlava Charles University,
Czech Republic
E-mail: malohlava@d3s.mff.cuni.cz

Mingshan Xie
Hainan University, China
E-mail: 283921977@qq.com

Mohamed Jmaiel
University of Sfax, Tunisia
E-mail: mohamed.jmaiel@enis.rnu.tn

Mohamed Rahmani
Dr Moulay Tahar University
of SAIDA, Algeria
E-mail: rahmanimed@yahoo.fr

Mona Abd Elghany
Arab Academy for Science
& Technology, Egypt
E-mail: mabdelghany2000@gmail.com

Nadia Bouassida
University of Sfax, Tunisia
E-mail: Nadia.Bouassida@isimsf.rnu.tn

Nermine Khalifa
Arab Academy for Science
& Technology, Egypt
E-mail: nerminek@gmail.com

Pacheco C
Technological University
of the Mixtec Region, Mexico
E-mail: leninca@mixteco.utm.mx

Petr Hnetynka
Michal Malohlava Charles University,
Czech Republic
E-mail: hnetynka@d3s.mff.cuni.cz

Raman Ramsin
Sharif University of Technology, Iran
E-mail: ramsin@sharif.edu

Reda Mohamed Hamou
Dr Moulay Tahar University
of SAIDA, Algeria
E-mail: hamoureda@yahoo.fr

Roger Y. Lee
Central Michigan University, USA
E-mail: lee1ry@cmich.edu

Rutvij Mehta
The University of Texas
at Dallas, USA
E-mail: rutvij.mehta@utdallas.edu

Sang-Seop Lim
JEJU National University,
Korea

Seok-Jun Ko
JEJU National University,
Korea

Seyyed Mohsen Hashemi
IAU University Science and Research
Branch, Iran
E-mail: hashemi@srbiau.ac.ir

Tomas Bures
Institute of Computer Science,
Czech Republic
E-mail: bures@d3s.mff.cuni.cz

Zahra Shakeri Hossein Abad
Sharif University of Technology, Iran
E-mail: z_shakeri@ce.sharif.edu

Identifying Critical Success Factors for Adopting CRM in Small: A Framework for Small and Medium Enterprises

I. Garcia, C. Pacheco, and A. Martinez

Abstract. Economic globalization and Information Technology development have intensified the competition among modern enterprises, independent of the trade and sector in which they develop. In this intense battle, the customer relationship has arisen as an important resource for establishing a competitive position against the rest of competitors. Customer Relationship Management (CRM), particularly, is a new technology, strategy and idea that provides overall guidelines and support for business in the construction, maintenance and management of customer relationships. However, CRM implementation into Mexican Small and Medium Enterprises (SMEs), is, until now, a field under continuous exploration which mainly generates recommendations about success and failure factors. However, there is evidence about the success of CRM in developing countries that enables countries to establish a comparative basis. This paper provides an exploratory study about CRM experiences in developing countries' SMEs, and presents a framework which implements the minimal requirements identified for this kind of enterprises in Mexico.

1 Introduction

Customer relationship management used to be easy. Traders knew their customers very well -what they bought in the past, what they will buy in the future, and their current and potential value as customers. This knowledge helped those traders to create highly efficient relationships with their customers. However, understanding of customer necessities was worn out because people began to move more rapidly,

I. Garcia · C. Pacheco
Technological University of the Mixtec Region
e-mail: ivan@mixteco.utm.mx, leninca@mixteco.utm.mx

A. Martinez
RagaSoft Inc.
e-mail: alberto@ragasoft.com.mx

R. Lee (Ed.): Software Engineering Research, Management and Appl. 2012, SCI 430, pp. 1–15.
springerlink.com

cities grew, companies became bigger, and the commercialization scope was enlarged. This development was unfortunate for both customers and enterprises in the same way. Nowadays, many companies are trying to get back to "those good old days" when they knew their customers perfectly through capturing abundant external and internal available data, analyzing those data to better understand the customer necessities, preferences and benefits, and using this knowledge in each communication with the customer. Thus, economic globalization, rapid technological development and the increasing power of customers cause business to face a more complex and dynamic environment for enterprises. Customer Relationship Management (CRM) represents the supporting tool for enterprise strategy to improve market competitiveness, while the main objective is to establish a correct relationship between businesses and customers; -that is, adequate products for adequate customers, at the right moment, by the appropriate means, and a win-win relationship related to the provision cost of product or service [4]. CRM extends the reach of commercialization using Information Technologies (IT) to assume the control of those aspects which require intensive work, and to make them feasible through a wide range of customers. The CRM current emphasis is lead by the demands of change in the enterprise environment, the availability of huge amounts of data, and the advances in IT. According to [70], IT is the most important instigator for CRM under a simple and intuitive concept: *"attract new customers, know them well, provide them with exceptional service, and anticipate their desires and necessities"*. When enterprises achieve these objectives, an increase in revenues and benefits will be obtained. However, CRM can mean different things for different people, and it can be implemented in many different ways. For some companies, CRM is related to the concept of creating offers based on customer behavior and demographic characteristics. For other people, CRM can mean a change in a website appearance based on customer profiles and information on their preferences. To include its diverse forms, the Data Warehousing Institute (DWI) defines CRM as follows: *"any application or initiative designed for supporting companies in the optimization of their interactions with customers, suppliers, or prospects through one or more contact points -as a support center, a vendor, a distributor, a warehouse, a branch, a Web site, or an e-mail- to acquire, maintain, and establish new customers"*. Some recent studies, for example [47; 16; 28] and [71], have shown that CRM gained impetus since 2000. Besides, a DWI study over 1500 companies found that since 2001, 91% of enterprises had planned CRM deployment in a very near future. Nevertheless, CRM was created for large enterprises which commonly have a large customer catalogue; but we consider that this is not a restriction for its adoption in Small and Medium Enterprises (SMEs), due to its size (related to staff and infrastructure) and budget which enable them to manage a respectable customer catalogue. Research on CRM in a SMEs context is under development. The study of Ozgener and Iraz [37], for example, is directly related to the applications of CRM in small environments. Other approaches have tried to study CRM from a more indirect way: as a topic related to electronic commerce [31; 54; 64]; as a topic related to relationship management [10; 68], or as a topic related to its adoption in SMEs [29]. Thus, the limited research that addresses CRM in SMEs implies that study is predisposed, mainly or in a large extent, by the experiences of large enterprises and, as a consequence, focused on their particular characteristics. An exploratory analysis of literature provides us with some evidence of research focused on SMEs particular characteristics. This literature discusses the existence of very important factors in SMEs for IT adoption (including CRM), for example:

1. Management dominion by enterprise owner [75; 31; 64; 35],
2. Independency commitment by the enterprise owner/manager [31; 10; 29],
3. Strategic myopia, ad hoc decisions, and constant inertia of politics [31; 64],
4. Commitment by the enterprise owner for customized customer relationships [29], and
5. Poor focus on longtime business relationships and orientation in local market [31; 64].

Concretely, SMEs have particular characteristics important for understanding their commercialization, and these characteristics should be addressed in research [18]. In other words, there is a necessity for developing a framework using empirical research over how SMEs manage relationships with their customers; and establishing the research basis over how to implement CRM in a similar context. The rest of the paper is organized as follows: Section 2 provides a background to CRM in developing countries and SMEs, and the motivation for this research. Moreover, this section summarizes the work related to improving the SMEs productivity using CRM-based efforts. Section 3 presents our CRM for SMEs framework focused on the adoption effort. Section 4 summarizes the main conclusions from this research.

2 CRM in Developing Countries

2.1 The Different Interpretations of CRM

Customers have been always the main preoccupation for all business in the world. From a marketing perspective, customers are considered the cornerstone for the organizations' activities. According to [40], this importance was reflected in the organizations' increased necessity for integrating customer knowledge with the objective for establishing close corporative relationships and societies with their customers. Thus, many efforts for tracing and categorizing CRM research have been performed. Romano and Fjermestad [50], for example, analyzed 362 journal articles and conference papers belonging to Electronic Customer Relationship Management (e-CRM). Papers were classified into five research areas: e-CRM markets, e-CRM business models, knowledge management using e-CRM, e-CRM technology, and e-CRM human factors. The study found that e-CRM technology was the most popular area, closely followed by e-CRM human factors and e-CRM markets. The e-knowledge management using CRM was the least popular research area. In 2005, Ngai extended the approach and analyzed 205 papers related to marketing, Information Systems (IS) and IT [34]. These papers were identified using functional categories as: CRM, marketing, sales, service and support, and IS and IT. The rest of the papers were considered as "general category". The Ngai study determined that the most common category was IS and IT, followed by the general category, and marketing, sales, and service and support categories. Paulissen et al. [41] analyzed 510 journal articles and conferences papers related to marketing and IS/IT disciplines. This analysis followed a classification scheme based on the different phases for a CRM lifecycle model: adoption, acquisition, implementation, usage and maintenance, evolution, and retirement. Publications that did not belong to any of these phases were classified as "general". The study found that the majority of papers belonged to the first phases of the lifecycle and the latter phases appeared less frequently. The three research studies, [50; 34] and [41], also determined that the number of publications related to CRM has substantially increased after the 90's. According to Wahlberg et al. [69], a fundamental problem in CRM research is that, at present, no common meaning of CRM exists. Rather, it appears to mean different things to different people. This has occurred mainly because different perspectives on the phenomenon are applied. For example:

- CRM is seen by some as a simple issue of integrating business process into an organization (i.e., [62; 65; 73; 47; 24; 43; 45]).
- Another perspective considers that CRM is an enterprise strategy focused on the customer (i.e., [10; 59; 36; 67]).
- The third major perspective considers that CRM is a matter of customer knowledge management (i.e., [76; 52; 57]). Closely related to this last perspective is seeing CRM as a matter of technology-enabled customer information management activities, including Strategic CRM, Analytical CRM, Operational CRM and Collaborative CRM (i.e., [25; 8; 66; 9]).

It is noteworthy that this last image has received a great deal of attention outside academic research. It is, for instance, a classification of CRM presented in Wikipedia, and also a classification advocated by the Meta Group. It is, however, not an image that is reflected in the attempts to describe the CRM research field mentioned above. Thus, according to Da Silva [13], this wide range of interpretations and differences in CRM perspectives could lead to the rise of the concept focused on two study fields: marketing and IT. From a technical point of view, Chen and Chen have defined CRM as: *"a methodology that heavily employs certain IT, such as databases and Internet, to leverage the effectiveness of the relationship marketing process"* [11]. In support of this definition, Ryals and Payne have referred to CRM as: *"to use the IT in implementing relationship marketing strategies"* [51]. To illustrate the emphasis on the technological element of CRM, Starkey and Woodcock defined it as: *"a process enhanced by IT that integrates the organizations' competences to deliver superior profitable customer value to existing and potential customers"* [63]. On the other hand, the wider definition of CRM concentrates on strategic orientation of CRM. Gray and Byun have defined CRM as: *"a primarily strategic business process issue rather than technology which consists of the following components: customer, relationship, and management"* [20]. Agreeing with the previous view, Brown referred to CRM as: *"a key competitive strategy that is needed to focus on customers' needs and to integrate an organizational customer-facing approach"* [7]. Reinartz and Chugh have provided support to the strategic definition of CRM. They have defined CRM as *"the strategic process of shaping the interactions between a company and its customers with the goal to maximize the lifetime value of customers for the company as well as to maximize satisfaction for the customer"* [46]. Despite the preferred approach of not considering CRM just a technology, the broad view of CRM also has some negative aspects. Such a broad view could result in a confusion that overlooks critical issues in implementing CRM such as the importance of IT and business process.

Analyzing previous examples of technical definitions for CRM, results in finding a lack of holistic overview of CRM which could lead to minimizing or even neglecting other aspects of CRM such as strategy and business processes. Consequently, such an approach will eventually contribute to the limitation of CRM outcomes. Thus, based on research of Payne [42], Payne and Frow [43], and Shang and Lin [56], the CRM definition for this research is: *"a strategic approach that integrates process, people, and technology cross functionally to understand an organization's customers, improve stakeholder value, and deliver profitable and long-term relationships with customers"*. Therefore, success factors for CRM lie in the scope of managing, integrating, and controlling CRM components.

2.2 CRM in Small

Wahlberg et al. developed a systematic review to describe the research field in CRM [69]. This review used the common variations of CRM (strategic, analytical, operational, and collaborative) to classify the found publications. This systematic review was performed in March 2007 and analyzed 468 research papers related to CRM. At the end of the 90's, the number of research papers focused on CRM as a specific topic was very low. The research findings of Romano and Fjermstad [50], Ngai [34], and Paulissen et al. [41] promoted CRM interest in enterprises and the number of research papers increased since 2001. However, CRM was relatively new in the marketing area and according to Wahlberg et al., in 2004 the number of research papers decreased significantly. The decrease in CRM articles is a finding that runs counter to Ngai's conclusion [34] that research on CRM will increase significantly in the future. However, when the distribution of articles was split up into the different branches of CRM (strategic, analytical, operational, and collaborative), it was evident that the decrease in numbers was very much due to a decrease in articles relating to Analytical CRM. This makes it rather difficult to interpret the 2005 decrease in the total number of articles on CRM. However, according to the experts, this decrease could be temporal. Furthermore, it is important to say that the decrease in 2006 could be caused by an 'editorial impetus' because the research papers used in the Wahlberg et al. systematic review were searched during March 2007. At that moment, all the research papers from 2006 could not be updated in the databases and thus the number of papers from 2006 will be uncertain. We complemented the Wahlberg et al. systematic review to discover a more detailed explanation about this tendency from 2005. We incorporated into the Wahlberg et al. research four full academic databases: Emerald, Science Direct, IEEE and ACM. The selection criteria for research papers was based on searching papers which include the "Customer Relationship Management" word in the title, in the keywords, or in the abstract. This search was performed in December 2010. Our systematic review followed the same research questions established by Wahlberg et al.:

- How has CRM research developed over the years?
- What topics are focused on in CRM research?

The obtained results demonstrate that in truth a reduction in the number of papers related to the CRM research existed. We did not find information that can explain this phenomenon, but it is possible that, as we mentioned, it is an editorial issue. Nevertheless, the 353 research papers found for the period 2006-2010 demonstrate that the tendency has exponentially increased, showing a total of 184 research papers for the closing of 2010 (without considering those papers that have not been updated in the data bases). According to our literature review, nowadays the main researched topics in the CRM field are summarized as Strategic, Analytical, Collaborative, Operational, Technical, Other, and Not applicable (these topics considered the 487 research papers analyzed by Wahlberg et al. for a total of 840 research papers from 1998 to 2010). According to 38 papers found through our systematic review, one of the new explored topics in the strategic CRM area is its relation with SMEs. Commonly, its adoption in this kind of enterprises is supported by the CRM software, explored in the CRM operational branch. Nowadays, the CRM adoption in SMEs is a research topic that has been in continuous and constant exploration (e.g., [37; 18; 74; 77; 60; 39]). However, the reduced number of research papers that address the use of CRM in SMEs indicates to us that this research area is highly influenced by the experiences of large enterprises and, consequently, with their particular characteristics. Moreover, research papers which focus on SMEs have discussed specific criteria important for CRM adoption. These criteria are summarized as follows:

1. The owner-manager dominance [10; 31; 35; 37].
2. The owner-manager commitment to independence [31; 54; 18].
3. Strategic myopia, ad hoc decisions and policy inertia. [31; 35].
4. Commitment to face-to-face and personalized customer relations [18].
5. Regarding selling as equal to marketing and sales people as the central market communication medium [35].
6. A focus on few and long term business relations and local market orientation [31; 35]. The focus on a few customers is particularly relevant for enterprises in the B2B sector - which is a very large sector of the economy in general.
7. Lack of marketing capabilities [35; 54].
8. Relatively low IT maturity and a lack of IT capabilities [35; 64].
9. And finally, a general lack of resources to finance IT investments [31; 37].

2.3 Success Factors to Adopt CRM in SMEs

According to Huotari and Wilson [23], Critical Success Factors (CSFs) arise from a methodology that focuses on identifying factors that are critical for an organization's success, as an absence of such factors could lead to its failure. The development of this methodology was based on the work by Daniel [14] and it was illustrated and gained more recognition by a paper by Rockart [48]. Although this methodology has been differently applied by researchers and practitioners, CSFs has received increased popularity in different fields of study and has been cited by many researchers such as Holland and Light [21], Slevin and Pinto [61], and Da Silva and Rahimi [13]. Thus, identifying CRM success factors is a critical issue to ensure successful implementation but more importantly it is required to link these factors to implementation process and manage them. In order to identify the CSFs for CRM implementation it is important to define what a "success factor" means. According to [38], *"the success factors could be defined as the generic ingredient that has to be the essential part of any successful CRM implementation"*. Besides, Esteves and Pastor [17] have identified that success factors are: *"the limited number of areas that when satisfactory will successfully improve the organization's competitive performance"*. However, according to Croteau and Li [12], there is no one universal procedure to collect and analyze CSFs. Hence, many research studies have applied diverse methods to identify these factors such as interviews followed by questionnaires and others. We performed an extensive literature review of over 18 studies (of the 38 papers previously identified) on CRM success factors to obtain a set of factors that could be considered for CRM implementation in SMEs (see Table 1). These success factors varied from wide and general to more specific and technical ones. Table 1 summarizes all the obtained CSFs applied to the SMEs context in two categories: organizational success factors and strategic/technical success factors. Organizational CSFs are related to those tasks that have to be established by the top management or senior executives; while strategic/technical CSFs are related to those tasks that have to be performed by the IT specialist or professionals guided by a manager who leads the established strategy. It is clear that the main objective of CRM research in the SMEs context is focused on establishing it in a simple and practical manner from the beginning, and supporting it later with systems integration and task specialization. However, the adoption of a CRM software tool in any SME strongly depends on the cultural factors and economics of the country; thus it is impossible to state that the CSFs identified in Table 1 could be adopted in any environment. However, they can be considered as *"base characteristics"* that can be useful to start the development of a software tool for supporting CRM implementation, and later be complemented with others that address the peculiarities of the industry where it will be applied.

Under this premise, CRM has been globally implemented by organizations from different economics and regions. Comparing the availability and richness of literature related to the CRM implementation in developed countries with the availability of such literature in developing countries, there is a gap that is favourable to the first ones. However, and according to [2], a lack of literature exists related to global CRM implementation and the impact of the cultural factors. In general, it is possible to affirm that a lack of research studies exists about CRM implementation in developing countries and, consequently, a lack of frameworks that manage the CRM implementation in these countries and that recognize, at the same time, their cultural

Table 1 Summary of previous studies on CRM success factors in SMEs

Organizational CSFs
System architecture [5]
Motivate staff [9]
Top management commitment; Customer information management; Customer service; Customer contact management; Information Systems integration [32]
Communication [13]
Organizational culture; Employees acceptance [15]
Customer involvement [38]
Willingness to share data [26]
CRM processes clearly defined; Technological formation [39]
Keep old customers, Identify new customers; Establish training mechanisms [33]
Strategic/technical CSFs
Align business and IT operations; Measure, monitor, and track [30]
Top management support [19; 12]
Gain board awareness of strategic potential of IT; Organize around customer [72]
Software customization [58]
Systems integration [11]
Customer information quality [49]
Understanding of customer behaviour; Extensive IT support [53]
Interdepartmental integration, Sales automation [77]
Customer satisfaction [74]

differences with the developed ones. Nevertheless, research by Wagner et al. [68], Al-Alawi [6], Ozgener and Iraz [37], Achuama and Usoro [1], Altomairi [3] and Sanzogni et al. [55], enable us to identify some common barriers that can affect CRM implementation in developing countries organizations. Thus, taking into account the literature analysis presented in this section, we present a framework that considers the CSFs (identified in Table 1) and the barriers in CRM adoption in developing countries, to support the implementation of a CRM approach in Mexican enterprises.

3 A CRM Framework for SMEs

Peppers and Rogers [44] considered that CRM was an enterprise strategy that can work as a norm based on three complementary perspectives: technology, business and customer. According to the experts, three major objectives for CRM have been found:

- Software -individual software tools for CRM which provide value to business.
- Infrastructure -data, software and hardware infrastructure that supports the CRM software tool and other future applications.
- Transformation -an organizational transformation can be originated through CRM efforts.

Producing an effective internal integration of information requires that enterprises overcome some traditional barriers in the IT successful adoption: willingness to share information, roles reorganization, and more [22]. Considering this logic, and according to Teo et al. [67], the CRM system architecture can be composed of three segments: operational CRM, collaborative CRM, and analytical CRM. However, the rigorous implementation of these three segments in a SME environment will hardly lead to success; due to the lack of staff skills, the limited budget in infrastructure for supporting independent systems, and the lack of timely information to make decisions related to any customer [27]. Considering that any SME is frequently composed of 25-500 employees, our framework promotes the implementation of two segments (operational CRM and analytical CRM), leaving the collaborative CRM for future research. Figure 1 shows the proposed technical architecture for CRM in SMEs environment. This architecture is modified from Goodhue et al. [19] architecture and it is defined for small contexts.

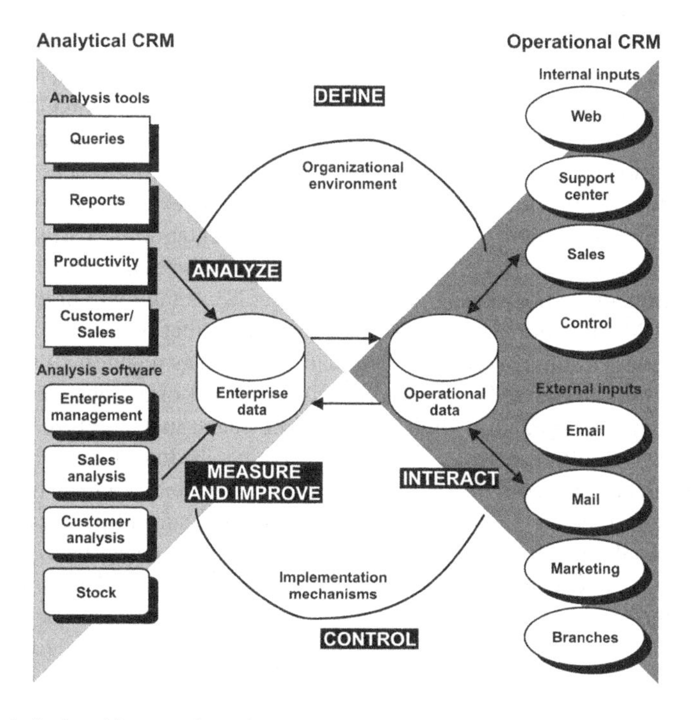

Fig. 1 Technical architecture for a CRM tool

As shown, the technical architecture for CRM in SMEs depends on software tools for performing both operational and analytical functions. In the operational context, data must be captured, integrated, and stored from all points of entry; including Web, branches, warehouses, and main office. This information can be complemented with other external entries. Current data can be maintained in an operational data repository that supports the operational tools, such as e-mail, marketing, and customer service. On the analytical side, a data repository usually stores the historical data that supports generic tasks, such as reports creation, queries performing, and online analytical process; as well as specific tasks, such as enterprise management, sales analysis, customer analysis, customer profit analysis, and stock analysis. This information is stored in the enterprise data repository. The most obvious aspects of this architecture are the diverse tasks which provide and use customer information, and the organizational and operational repositories which store all the information needed to enhance customer relationships. Two more important components for a successful CRM effort in SMEs are implicit in Figure 1: the needed infrastructure to make data distribution possible, and the organizational transformation needed to completely take advantage of the capabilities that CRM provides. Thus, the renovated focus for providing a SME with the capability to comprehensively accede to customer information (and as a consequence to the obtained profits by analyzing such information) probably introduces new ways that the SME could use to interact with them. However, these benefits and possibilities will not have an impact if the SME does not change the way to do business to take advantage of them. Moreover, a CRM strategy can imply an important SME transformation. Once the CRM technical architecture for a tool in a small context has been defined, it is necessary to define the manner of integrating it into a particular SME. Unlike the inexistence of frameworks to define architecture for CRM tools in SME context; research by [38; 77; 74; 39] provides information related to CRM implementation in SMEs. Although these studies are not punctual guidelines, they provide solid help for searching for a successful implementation. Using the research by [38] as the main orientation, we propose a CRM integration method focused on SMEs that is oriented toward organizational process improvement. Our method establishes a generic improvement model (according to the SME characteristics) composed of 5 stages and 14 steps (Establish the project goal and align it with the business goal, Analyze the current process, Select an efficient team, Develop a project schedule, Determine what to measure, Collect and record the data, Determine the baseline, Analyze problems in current process, Identify a CRM process to redefine the current process, Design a CRM tool, Install the CRM tool and train users, Manage changes, Measure defined values, and Document the learned lesson and identify new CRM goals). According to a generic process improvement model, the method stages can be defined as follows:

- *Define:* This stage establishes the implementation context, that is, it determines the real necessity to implement a CRM strategy in such a manner that it allows aligning the searched objective with the business objectives. It is important to say that it is in this stage where top management commitment must become patent.
- *Measure:* This stage identifies measurable attributes that allow verifying the effectiveness of the implementation through the definition of baseline. Commonly for a SME, the productivity and sales volume indicators can be used as successful metrics.

- *Analyze:* This stage performs an analysis over the current process with the aim to determine strengths and weaknesses and to establish a CRM process according to the SME real necessities. Finally, the design of the CRM tool that will be implemented in later stages begins.
- *Improve:* This stage focuses on installing the CRM tool developed in the previous stage and managing the changes or adjustments that the SME marks during the training of staff.
- *Control:* This stage concludes the implementation cycle with the collection of data and learned lessons. The attributes measurement defined in step 7 will allow to determine the implementation success.

4 Conclusions and Future Research

The contributions made by this research emerge from different parts of this paper. Such contributions include the contextual information provided in Sections 1 and 2, and the development of the conceptual framework for CRM small-scale implementation. The work presented in this paper has made an alternative contribution to the field of CRM implementation with a concentration in CRM's successful implementation in SMEs. Through evaluating the conceptual framework in a developing country, factors that affect CRM success were monitored to conduct any required modification for the framework. In our case, the case study, which was conducted in Mexico as an exemplar of developing countries, revealed an effect of culture on successful CRM implementation. However, the development of a CRM tool to support the CRM implementation will reduce this cultural effect. Such effect of technology on CRM was stated by Ward: *"IT is the most important instigator for CRM under a simple and intuitive concept: attract new customers, know them well, provide to them an exceptional service, and anticipate their desires and necessities"*. The proposed conceptual framework integrates the CRM CSFs associated with causes of CRM failure and that are accepted widely by the literature. Additionally, our framework also considered the strategic nature of CRM by integrating CRM major components and emphasizing the need of CRM strategy development. Moreover, the conceptual framework includes a CRM Tool to help small organizations to increase revenues and benefits.

References

1. Achuama, P., Usoro, A.: Dancing with Stars: CRM and SMEs in Developing Countries. In: Proc. of the First International Business Conference, Dearborn, Detroit Metro, August 7-9 (2008)
2. Ali, M., Brooks, L., Alshawi, S., Papazafeiropoulou, A.: Cultural Dimensions and CRM Systems Implications: A Preliminary Framework. In: Proc. of the 12th Americas Conference on Information Systems, AMCIS 2006, Acapulco, Mexico (August 2006)
3. Almotairi, M.: Evaluation of the Implementation of CRM in Developing Countries. Doctoral Thesis. Brunel University - Brunel Business School, UK (2008)

4. Alonso-Mendo, F., Fitzgerald, G.: A multidimensional framework for SME e-business progression. Journal of Enterprise Information Management 18(6), 678–696 (2005)
5. Alt, R., Puschmann, T.: Successful Practices in Customer Relationship Management. In: Proc. of the 37th Hawaii International Conference on System Science, HICSS 2004, pp. 70–78. IEEE Computer Society (2004)
6. Al-Alawi, A.: Customer Relationship Management in the Kingdom of Bahrain. Issues in Information Systems 5(2), 380–385 (2004)
7. Brown, S.A.: A case study on CRM and mass customization. In: Brown, S.A. (ed.) Customer Relationship Management: A Strategic Imperative in the World of E-business, pp. 41–53. Wiley, Toronto (2002)
8. Buttle, F.: Customer Relationship Management Concepts and Tools. Butterworth Heineman, Oxford (2004)
9. Chalmeta, R.: Methodology for customer relationship management. The Journal of Systems and Software 79(7), 1015–1024 (2006)
10. Chen, I., Popovich, K.: Understanding customer relationship management (CRM): People, process and technology. Business Process Management Journal 9(5), 672–688 (2003)
11. Chen, Q., Chen, H.: Exploring the success factors of e-CRM strategies in practice. Journal of Database Marketing and Customer Strategy Management 11(4), 333–343 (2004)
12. Croteau, A.-M., Li, P.: Critical Success Factors of CRM Technological Initiatives. Canadian Journal of Administrative Sciences 20(1), 21–34 (2003)
13. Da Silva, R., Rahimi, I.: A Critical Success Factors model for CRM Implementation. International Journal of Electronic Relationship Management 1(1), 3–15 (2007)
14. Daniel, D.R.: Management Information Crisis. Harvard Business Review, 111–121 (September-October 1961)
15. Eid, R.: Towards a Successful CRM Implementation in Banks: An Integrated Model. The Service Industries Journal 27(8), 1021–1039 (2007)
16. El Sawah, S., Tharwat, A., Rasmy, M.: A quantitative model to predict the Egyptian ERP implementation success index. Business Process Management Journal 14(3), 288–306 (2008)
17. Esteves, J., Pastor, J.: Enterprise resource planning systems research: An annotated bibliography. Communication of AIS 7(8), 51–52 (2001)
18. Fink, D., Disterer, G.: International case studies - To what extent is ICT infused into the operations of SMEs? Journal of Enterprise Information Management 19(6), 608–624 (2006)
19. Goodhue, D., Wixom, B., Watson, H.: Realizing Business Benefits through CRM: Hitting the Right Target in the Right Way. MIS Quarterly Executive, 79–94 (June 2002)
20. Gray, P., Byun, J.: Customer Relationship Management. Centre for Research on Information Technology and Organizations, Version 3-6, University of California (2001)
21. Holland, C., Light, B.: A Critical Success Factors Model for ERP Implementation. IEEE Software 16(3), 30–36 (1999)
22. Hu, S.R.: Association studies of core capacity and customer relationship management. Economic Research Tribune (2), 55–57 (2006)
23. Huotari, M., Wilson, T.: Determining organizational information needs: the Critical Success Factors approach. Information Research 6(3) (2001)
24. Jain, S.: CRM shifts the paradigm. Journal of Strategic Marketing 13(4), 275–291 (2005)
25. Karimi, J., Somers, T.M., Gupta, Y.P.: Impact of Information Technology Management Practices on Customer Service. Journal of Management Information Systems 17(4), 125–158 (2001)

26. King, S., Burgess, T.: Understanding Success and Failure in Customer Relationship Management. Industrial Marketing Management 34(4), 421–431 (2008)
27. Kujawiak, L., Sakowicz, B., Chlapinski, J., Mazur, P.: Small and medium enterprises supporting system based on integration of CMS and CRM solutions using .NET framework. In: Proc. of the 10th International Conference - The Experience of Designing and Application of CAD Systems in Microelectronics, CADSM 2009, pp. 439–442. IEEE Computer Society (2009)
28. Lei, Y., Yang, H.: The Application Research of Web 2.0 in Customer Relationship Management. In: Proc. of the 2010 Conference on e-business and e-government, pp. 3153–3155. IEEE Computer Society (2010)
29. Li, B., Liang, Q.: Research on Customer Relationship Management System based on SOA. In: Proc. of the First International Workshop on Education Technology and Computer Science, ETCS, pp. 508–511. IEEE Computer Society (2009)
30. Mankoff, S.: Ten Critical Success Factors for CRM: Lessons Learned from Successful Implementations. Siebel System, White Paper (2001)
31. Manuel, P., Demirel, O., Gorener, R.: Application of e-Commerce for SMEs by using NLP principles. In: Proc. of the 2003 Engineering Management Conference, IEMC 2003, Managing Technologically Driven Organizations: The Human Side of Innovation and Change, pp. 470–473 (2003)
32. Mendoza, L., Marius, A., Perez, M., Griman, A.: Critical success factors for a customer strategy. Information Software Technology 49, 913–945 (2007)
33. Min, L., Hui, Z., Xuwen, G.: Research on customer relationship management for small and medium-sized enterprise based on implementation strategies. In: Proc. of the 2011 International Conference on E -Business and E -Government, ICEE 2011, pp. 1–4. IEEE Computer Society (2011)
34. Ngai, E.: Customer relationship management research (1992-2002). An academic literature review and classification. Marketing Intelligence and Planning 23(6), 582–605 (2005)
35. O'Toole, T.E.: Relationships - emergence and the small firm. Marketing Intelligence and Planning 21(2), 115–122 (2003)
36. Osarenkhoe, A.: Customer-Centric Strategy: A longitudinal study of implementation of a customer relationship management solution. International Journal of Technology Marketing 1(2), 115–143 (2006)
37. Ozgener, S., Iraz, R.: Customer relationship management in small-medium enterprises: The case of Turkish tourism industry. Tourism Management 27(6), 1356–1363 (2006)
38. Pan, Z., Ryu, H., Baik, J.: A Case study: CRM Adoption Success Factors Analysis and Six Sigma DMAIC Application. In: Proc. of the 5th ACIS International Conference on Software Engineering Research, Management and Applications, SERA 2007, pp. 828–835. IEEE Computer Society (2007)
39. Pan, W., Wang, D.-S.: The Development and Application of Customer Relationship Management System. In: Proc. of the International Conference on Intelligent Computation Technology and Automation, ICICTA 2010, pp. 788–791. IEEE Computer Society (2010)
40. Parvatiyar, A., Sheth, J.N.: Customer relationship management: Emerging practice, process, and discipline. Journal of Economic and Social Research 3(2), 1–34 (2002)
41. Paulissen, K., Mills, K., Brengman, M., Fjermestad, J., Romano, N.: Voids in the Current CRM Literature: Academic Literature Review and Classification (2000-2005). In: Proc. of the 40th Annual Hawaii International Convergence on System Sciences, HICSS 2007, pp. 150–160. IEEE Computer Society (2007)

42. Payne, A., Frow, P.: The role of multichannel integration in customer relationship management. Industrial Marketing Management 33(6), 527–538 (2004)
43. Payne, A., Frow, P.: A Strategic Framework for Customer Relationship Management. Journal of Marketing 69(4), 167–176 (2005)
44. Peppers, D., Rogers, M.: The One to One Future, Division of Bantam, DoubleDay. Dell Publishing Group, New York (1993)
45. Plakoyiannaki, E., Saren, M.: Time and the customer relationship management process: conceptual and methodological insights. Journal of Business and Industrial Marketing 21(4), 218–230 (2006)
46. Reinartz, W., Chugh, P.: Learning from experience: making CRM a success at last. International Journal of Call Centre Management 4(3), 207–219 (2002)
47. Reinartz, W., Krafft, M., Hoyer, W.: The Customer Relationship Management Process: Its Measurement and Impact on Performance. Journal of Marketing Research 41(3), 293–305 (2004)
48. Rockart, J.: Chief Executives Define Their Own Information Needs. Harvard Business Review, 81–92 (March/April 1979)
49. Roh, T., Ahn, C., Han, I.: The priority factor model for customer relationship management system success. Expert Systems with Applications 28(4), 641–654 (2005)
50. Romano, N., Fjermestad, J.: Electronic Commerce Customer Relationship Management: A Research Agenda. Information Technology and Management 4(2-3), 233–258 (2002)
51. Ryals, L., Payne, A.: Customer relationship management in financial services: towards information-enabled relationship marketing. Journal of Strategic Marketing 9(1), 3–27 (2001)
52. Salahi, A., Fakhry, H.: Study and development of CRM information system for small and medium businesses. In: Proc. of the 2004 International Conference on Information and Communications Technologies: From Theory to Applications, ICTTA 2004, pp. 503–504. IEEE Computer Society (2004)
53. Saloman, H., Dous, M., Kolbe, L., Brenner, W.: Customer Relationship Management Survey, Status Quo and Future Challenges. Institute of Information Management. University of St. Gallen (2005)
54. Sand, M.: Integrating the Web an e-mail into a push-pull strategy. Qualitative Market Research: an International Journal 6(1), 27–37 (2003)
55. Sanzogni, L., Whungsuriya, J., Heather, G.: Corporate Struggle with ICT in Thailand: A Case Study. The Electronic Journal of Information System in Developing Countries 34(3), 1–15 (2008)
56. Shang, S., Lin, J.: A Model for Understanding the Market-orientation Effects of CRM on the Organizational Processes. In: Proc. of the Eleventh Americas Conference on Information Systems, AMCIS 2005, AIS Electronic Library (2005)
57. Shen, Y., Song Shun, L., Li Shou, W.: The Design and Implement of CRM Data Mining System for Medium-Small Enterprises Based on Weka. In: Proc. of the International Forum on Information Technology and Applications, IFITA 2009, pp. 596–599. IEEE Computer Society (2009)
58. Siebel Systems. Critical Success Factors for Phased CRM Implementations. Siebel White Papers (2004)
59. Sin, L., Tse, A., Yim, F.: CRM: conceptualization and scale development. European Journal of Marketing 39(11-12), 1264–1290 (2005)
60. Siyu, X., Genguo, C.: Application Research of CRM based on SaaS. In: Proc. of the 2010 Conference on E-Business and E-Government, ICEE 2010, pp. 3090–3092. IEEE Computer Society (2010)

61. Slevin, D., Pinto, J.: Balancing strategy and tactics in project implementation. Sloan Management Review 29(1), 33–41 (1987)
62. Srivastava, R., Shervani, T.A., Fahey, L.: Marketing, business processes, and shareholder value: an organizationally embedded view of marketing activities and the discipline of marketing. Journal of Marketing 63, 168–179 (1999)
63. Starkey, M., Woodcock, N.: CRM systems: Necessary but not sufficient. REAP the benefits of customer management. Journal of Database Marketing 9(3), 267–275 (2002)
64. Subba Rao, S., Metts, G.: Electronic commerce development in small and medium sized enterprises. Business Process Management Journal 9(1), 11–32 (2003)
65. Swift, R.S.: Accelerating Customer Relationships using CRM and Relationship Technologies. Prentice Hall, Englewood Cliffs (2001)
66. Tanner, J.F.: CRM in sales-intensive organizations; a review and future directions. Journal of Personal Selling and Sales Management 25(2), 169–180 (2005)
67. Teo, T., Devadoss, P., Pan, S.: Towards a holistic perspective of customer relationship management (CRM) implementation: A case study of the Housing and Development Board, Singapore. Decision Support Systems 42(3), 1613–1627 (2006)
68. Wagner, C., Cheung, K., Lee, F., Ip, R.: Enhancing e-Government in Developing Countries: Managing Knowledge through Virtual Communities. The Electronic Journal of Information Systems in Developing Countries 14, 1–20 (2003)
69. Wahlberg, O., Strandberg, C., Sundberg, H., Sandberg, K.: Trends, topics and under-researched areas in CRM research. International Journal of Public Information Systems 3, 191–208 (2009)
70. Ward, J., Peppard, J.: Strategic Planning for Information Systems, 3rd edn. Wiley Series in Information Systems (2005)
71. Wei-wei, P., Dong-sheng, W.: The Development and Application of Customer Relationship Management System. In: Proc. of the 2010 International Conference on Intelligent Computation Technology and Automation, ICICTA 2010, pp. 788–791. IEEE Computer Society (2010)
72. Wilson, H., Daniel, E., McDonald, M.: Factors for Success in Customer Relationship Management (CRM) Systems. Journal of Marketing Management 18(1-2), 193–219 (2002)
73. Winer, R.S.: A framework for customer relationship management. California Management Review 43(4), 89–105 (2001)
74. Wu, Y., Zhang, S.: System Design of Customer Relation Management System of Small and Medium-sized Wholesale and Retail Enterprises. In: Proc. of the International Symposium on Knowledge Acquisition and Modeling, KAM 2008, pp. 789–793. IEEE Computer Society (2008)
75. Yu, L.: Successful Customer-Relationship Management. Sloan Management Review 42(4), 18–19 (2001)
76. Zablah, A.R., Bellenger, D.N., Johnston, W.J.: An evaluation of divergent perspectives on customer relationship management: Towards a common understanding of an emerging phenomenon. Industrial Marketing Management 33(6), 475–489 (2004)
77. Zheng, L., Pan, T., Ren, G., Fang, C.: Development and Implementation of ERP/CRM System Based on Open Source Software to Small and Medium-sized Enterprise in China. In: Proc. of the International Conference on Intelligent Computation Technology and Automation, ICICTA 2008, pp. 725–729. IEEE Computer Society (2008)

A New Biomimetic Approach Based on Social Spiders for Clustering of Text

Reda Mohamed Hamou, Abdelmalek Amine, and Mohamed Rahmani

Abstract. View the explosion of data volume and high circulating on the web (satellite data, genomic data ...) the classification of the data (data mining technique) is required. The clustering was performed by a method based bio (social spiders) because there is currently no method of learning that can almost directly represent unstructured data (text). Thus, to make a good data classification must be a good representation of the data. The representation of these data is performed by a vector whose components are derived from the overall weight of the corpus used (TF-IDF). A language-independent method was used to represent text documents is that of n-grams characters and words. Several similarity measures have been tested. To validate the classification we used a measure of assessment based on recall and precision (f-measure).

1 Introduction and Background

Given the problems faced by the supervised classification such as the need of many human, the initial classification should be reviewed when the number of documents increases, according to [2] where about 80% of documents are in text format. This huge volume of unstructured or semi structured gives rise to an act to find relevant information more difficult to achieve, that creates a problem known as the problem of information overload [1]. The techniques and tools for knowledge discovery in texts (KDT) [3] or simply text mining [2] are being developed to address this problem. One such technique is clustering, a technique for grouping similar documents of a given collection by helping to understand its contents [4, 5]. One of his goals is to similar documents in the same group and placing documents in various different groups. The assumption is that through a process of clustering, similar objects

Reda Mohamed Hamou · Abdelmalek Amine · Mohamed Rahmani
Dr. Moulay Tahar University of SAIDA Mathematics and
Computer Science Department Saida, Algeria
e-mail: hamoureda@yahoo.fr, amine_abd1@yahoo.fr,
rahmanimed@yahoo.fr

R. Lee (Ed.): Software Engineering Research, Management and Appl. 2012, SCI 430, pp. 17–30.
springerlink.com

remain in the same group based on the attributes they have in common. This assumption is known as the hypothesis of cluster described by [6]. In this paper, we introduced a new model from nature in this case the social spiders to solve a critical problem of data mining or text mining more precisely what the clustering of data by which we hope to contribute about solving this problem because the grouping or clustering of textual documents, especially Web pages, is one of the challenges of current research.

2 State of the Art

A well designed clustering algorithm generally follows the four phases of design: data representation, modeling, optimization, and validation [7]. Phase representation of the data structures predetermine what kind of cluster can be found in the data. Based on data representation, the modeling phase defines clusters and the criteria that separate the desired group structures of these unwanted or unfavorable. In this phase, a measure of quality, which can be either optimized or approximate when searching hidden structures in the data is produced.

3 Data Representation

Learning algorithms can not deal directly with unstructured data such as text for this you have to go through a stage of indexing which is simply a representation of the text in numeric vector. But this step is very delicate and very important because a poor representation of data results in a misclassification of such data. The digital components of the vector associated with the text are the words of the text document that is associated with a weight. In this way learning algorithms directly uses the vectors that represent text. The choice of terms is to convert each document into a vector v_i d_i $(W_{1J}, w_{2j}, ..., W_{|T|j})$ where T is the number of all terms that appear at least once in the corpus. W_{KI} weight indicates the importance of the term t_k in document d_i. In our study the representations used are those bags of words that represent the simplest representation of textual records and that was introduced in the vector model and the n-grams that operate independently of the language of the corpus used. The principle of representation of the first approach is simply to turn each text into a vector where each component represents a word. With this approach, documents are represented by vectors of very large which makes the learning algorithms very difficult to use. For this we had to make a reduction in size as we will explain in the next section. Once the choice of components of the vector representing a text d_i is made, it must be consolidated for each coordinate of the vector v_i by calculating the weight of each term W_{KI} as follows:

- Binary: "0" to indicate that the term does not exist in the document and "1" to indicate that the term exists in the document.
- For instance: W_{KI} equal to the number of times the word appears in the document. Coding methods are based on two observations:

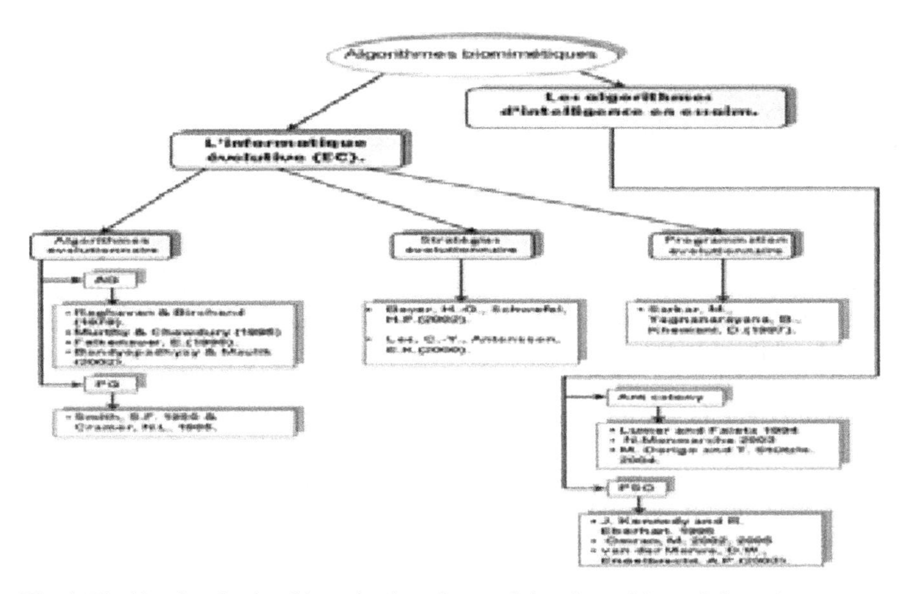

Fig. 1 The biomimetic algorithms that have been solving the problem of clustering

- Plus the word is common in a text, it is more related to the topic of this text.
- Plus the word is common in a corpus, unless it will be used to differentiate between documents.

$TF-IDF$ weighting ("Term Frequency-Inverse Document Frequency") is used to calculate the weight of a term. This gives more weight to words that appear often within the same text, which corresponds to the first observation. But its peculiarity is that it also gives less weight to words that belong to several documents and corresponding to the second observation. Encoding $TF-IDF$ does not correct the document length, coding for this TFC is similar to that of $TF*IDF$, but it fixes the lengths of the texts by cosine normalization, not favor longer documents. It is formalized as follows:

$$TFC(t_k,d_j) = -\frac{TF\times IDF(t_k,d_j)}{\sqrt{\sum_{s=1}^{r}(TF\times IDF(t_k,d_j))}}$$

3.1 Dimension Reduction

The data size is calculated from the number of variables (term) and the number of examples (text). In Text Mining, variables can reach hundreds of thousands, which can be a problem in the exploration and analysis. The reduction in size solves this problem and its objective is to select or retrieve an optimal subset of relevant features on a standard set. The selection of this subset of features leads to the elimination of redundant and irrelevant information and therefore it allows the reduction of dimension. Indeed, the main objectives of reduced dimension are:

- To facilitate the visualization and understanding of data
- Reduce storage space necessary
- Reduce the time to learn and use,
- Identify the relevant factors.

The dimension reduction techniques are classified into two broad categories: those based on the selection of attributes "feature selection", which retains only the attributes (terms) considered useful, according to some evaluation function. Others are rejected and those based on the extraction of attributes "feature extraction", which creates new attributes, or by combinations or transformations from the attributes (terms) of origin. In our study we opted for a technique based on the selection of attributes that represents the chi^2. It assesses the lack of independence between a word and a. It uses the same concepts of co-occurrence word / category as the mutual information, but an important difference is that it is subject to a standard that makes them comparable terms. She still loses relevance for infrequent terms. It is formalized as follows:

$$X^2(t_k,c_i) = \frac{|T_r|[P(t_k,c_i).P(\overline{t_k},\overline{c_i}) - P(t_k,c_i).P(\overline{t_k},c_i)]^2}{P(t_k).P(\overline{t_k}).P(c_i).P(\overline{c_i})}$$

3.2 *Similarity Matrix*

After representing our text documents into digital data and prior to the classification, by our new approach based on social spiders, a very important concept in our context is the similarity. Evaluate similarities between textual entities is one of the central problems in several disciplines such as textual data analysis, information retrieval and knowledge extraction from textual data (text mining). To produce the structures that will be used to represent texts in the calculation of similarities, textual data must first be broken down into simple tokens. The objective here is to see whether two documents are close to each other or, more precisely if they are similar. It is said that two documents are close to each other if the distance between them is low. It is said that two documents are similar if they are similar. If the distance between D_1 and D_2 is low then their similarity is high.

3.3 *Different Similarity Distance*

3.3.1 Jaccard Index

Jaccard index is calculated as follows:

$$s_j = \frac{m_c}{m_1 + m_2 - m_c}$$

m_c = number of words in common = shared vocabulary
m_1 = size of the lexicon of D_1 (ie. number of different words in D_1)
m_2 = size of the lexicon document D_2
The vectors used in this calculation of the similarity index of Jaccard are based on the presence / absence of words that can be characterized by a 0 (no) or a 1 (presence).

3.3.2 Minkowsky Distance of Order r

$$d_{min}(x,y) = (\sum_{j=1}^{d} |x_j - y_j|^r)^{\frac{1}{r}}, r \geq 1$$

When r = 1 is referred to as Manhattan distance or city block and r = 2 is the Euclidean distance.

3.3.3 Cosine Distance

For this measurement, we use the full vector representation, that is to say, with word frequency. It is defined as:

$$cos\alpha = cos(V_1, V_2) = \frac{V_1.V_2}{||V_1||.||V_2||}$$

3.3.4 Mahalanobis Distance

The Mahalanobis distance can also be defined as the dissimilarity measure between two vectors (documents) x and y the same distribution with a covariance matrix Σ:

$$d(\vec{x}, \vec{y}) = \sqrt{(\vec{x} - \vec{y})^T \Sigma^{-1} (\vec{x} - \vec{y})}$$

If the covariance matrix is the identity matrix, then the distance is the same as the Euclidean distance. If the covariance matrix is diagonal, it is called normalized Euclidean distance:

$$d(\vec{x}, \vec{y}) = \sqrt{\sum_{i=1}^{p} \frac{(x_i - y_i)^2}{\sigma_i^2}}$$

σ_i which is the standard deviation of x_i on the data set.

3.3.5 Chebyshev Distance

The distance is a Chebyshev metric defined on the vector space where the distance between two vectors is the largest of their differences in any dimension of the same rank. It is defined as:

$$\lim_{p \to \infty} \sqrt{\sum_{i=1}^{n} |x_i - y_i|^p} = \sup_{1 \leq i \leq n} |x_i - y_i|$$

Given the varying distances of similarity, it is difficult to say it and better than the other, this comparison will be made after an experiment and we can see beyond the quality of each distance.

4 Our Approach

Human beings have the natural ability to do the clustering of objects. Given a box full of marbles of four different colors: red, green, blue and yellow, even a child can separate these beads into four groups according to their colors. However, this type of problem is quite difficult to solve for a computer. The major obstacle in this task is that the brain is much less well understood by scientists saw the mechanisms by which it stores huge amounts of information processing at speeds of lightning and the deduction rules fitted with direction.

The behavior of an ant, bee, termite, wasp or spider is often too simple, but their collective behavior and social development is of paramount importance. By focusing on the animal species, particularly mammals including lions who enjoy a social life, perhaps for self existence in old age, particularly when they are injured. Collective behavior and social life of the creatures have motivated researchers to undertake the study of what is now known as the "Swarm Intelligence". It is in the 80's that work has been done to study the behavior of living things different versatile and especially the social insects. Efforts to mimic such behavior through computer simulation finally resulted in the fascinating field of SI. Systems or models based on Swarm Intelligence generally consist of a population of simple agents (an entity capable of performing certain operations) that interact locally with each other and their environment. Local interactions between such agents often lead to the emergence of global behavior. Among these algorithms include the most common in the literature and in this case the ant colony algorithm (ACO) algorithms and particle swarm (PSO). Social spiders that are detailed below represent our model in resolving the problem of clustering of texts.

4.1 Social Spiders

There is only fifteen species of social spiders among the 35,000 known species [Avils 1997]. All these species, except one whose collaborative nature is not clear, weave the Commons, which emphasizes the role of silk in the achievement of collective tasks. A spider is made of an internal state and a set of features. The internal state includes a current position and the last position where the spider had woven. The features of the spider social spiders have three features:

a)- Movement
b)- Weaving of a wire between two stakes,
c)- Return to the last position which was held weaving.

The life cycle of a spider is to run successively each of the features mentioned in the preceding order.

Movement: The movement allows the spider to move in the environment. This feature is to update the position of the spider from its neighbors. The choice of the new position is primarily dependent on the number of colonies present in the environment.
Weaving: Weaving creates a link between the current position and the last position where a wire was woven.
Backspace: The backspace is to return to the last position where a wire was woven.

4.2 The Behavioral Model Construction

To solve the task of clustering of data by the behavioral model of social spiders we need to describe the build environment of spiders, agents and their behavior and the dynamics of the system.

4.2.1 The Environment

It is a square grid where each box corresponds to a stake, they can be of different heights, representing the roughness of the support plant. The stakes can be linked together by the son of silk still attached to the top. It is noteworthy that the height of the posts is not used in the behavior for the simulation (3D is only used for visualization).

4.2.2 The Agents and Their Behaviors

Agents correspond to the spiders. They are described by two types of behaviors that can each be performed in a simulation cycle: This is the moving and installation of a wire. The spider moves always leaving behind a wire that is attached randomly. When attached, the wire is the shortest segment between the last spike and the installation of the post current. To move from a given pole, a spider has three possible choices (not mutually exclusive):

a)- Go on a pole adjacent;
b)- Go to a post following a thread that has woven itself;
c)- Go to a post following a thread woven by one of its congeners.

We associate with every choice a weighting constant with respect to the adjacent poles, set for the last two respectively by what we call factor of attraction for its silk and coefficient of attraction for other silk. Each rod is then weighted relatively accessible ways to access them, this weight when normalized gives the probability for the spider to access this box. This is called non-deterministic or probabilistic model.

4.2.3 The Dynamics of the System

It is built on the principle of stigmergy: the behavior of agents have effects on the environment (laying son), in return behavior is influenced by the context of the agent. Thus, agents are trapped by their silk in relation to the attraction it exerts.

4.3 Clustering by Social Spiders

In our study the construction of the canvas is as follows:

4.3.1 Environment

Grid rod $(n \times n)/n$ = sqrt (number of document) where each pole represents a document. The stakes are the same size. A woven wire means that the document is similar to the arrival of paper from the spider.

4.3.2 Behaviour

Displacement: The displacement of the spider is incremental and random. Weaving: If a document is similar to the first document in the cluster (centroid sort through a picket) while a thread is woven with the last document that was added to the cluster Communication: There is no communication between agents because there is a single spider. A "path" is a cluster and the road connecting rods are the documents of the cluster.

4.4 Experimentation

Before presenting the results of clustering, we detail our data sources that represent the benchmark Reuters.

4.4.1 Reuters 21.758

We used in our experiments the Reuters 21578 corpus which is a database of 21,578 new informational text documents in English.

4.4.2 Validation Tools Used

After experimenting with our algorithm on the Reuters 21578 corpus corpus we obtained the following results in terms of number of classes and purity of clusters. In terms of pure class and error rate of classification we used an evaluation measure that is the f-measure based on two concepts: the recall and precision defined as follows:

$$precision(i,k) = \frac{N_{ik}}{N_k}$$

$$recall(i,k) = \frac{N_{ik}}{N_{c_i}}$$

Where N is the total number of documents, i is the number of classes (predefined), K is the number of classes of the unsupervised classification, N_{C_I} is the number of documents in class i, N_k is the number of documents in the cluster C_k, N_{ik} is the number of documents of class i in cluster C_k. The f-measure is calculated on the partition P as follows:

$$F(p) = \sum \frac{N_{c_i}}{N} \max \frac{(1+\beta) \times recall(i,k) \times precision(i,k)}{\beta \times recall(i,k) + precision(i,k)}$$

β is used to control the important reminder about the accuracy, typically set to 1. The partition P is the expected solution is one that maximizes the F-measure associated. (In our study P is the partition corresponding to the class of the classification results by the method of 2D cellular automata for the number of related documents).

4.4.3 Results

After experimentation we have grouped the results in the tables below.

Table 1 Results of classification for social spiders, representation of data reduction dimension (CHI_2) before coding TFC

				N-grams character						1-gram word		
	2			3			4			1		
Distance	# Class	Learning time(S)	F-measure	# Class	Learning time(S)	F-measure	# Class	Learning time(S)	F-measure	# Class	Learning time(S)	F-measure
Jaccard	66	4"862	0.232	61	4"263	0.233	80	6"484	0.235	63	4"383	0.240
Cosine	13	1"687	0.414	34	3"313	0.217	99	6"765	0.264	49	4"082	0.231
Euclidian	11	0"891	0.347	37	2"963	0.313	32	3"087	0.245	9	0"687	0.362
Manhattan	32	3"087	0.319	45	3"198	0.350	46	3"331	0.354	35	3"315	0.314
Minkowsky3	13	1"687	0.299	32	3"087	0.380	36	2"748	0.371	38	3"540	0.337
Mahalanobis	27	2"703	0.562	17	1"343	0.533	19	1"453	0.549	17	1"343	0.534

Table 2 Results of classification for social spiders representation of data without reduction coding dimension after TFC

				N-grams character						1-gram word		
	2			3			4			1		
Distance	# Class	Learning time(S)	F-measure	# Class	Learning time(S)	F-measure	# Class	Learning time(S)	F-measure	# Class	Learning time(S)	F-measure
Jaccard	61	4"750	0.236	32	3"087	0.309	88	6"125	0.299	70	5"499	0.231
Cosine	20	1"516	0.440	35	3"313	0.368	42	3"479	0.269	24	1"928	0.217
Euclidian	13	1"094	0.205	34	2"796	0.368	20	1"516	0.365	22	1"828	0.369
Manhattan	24	1"908	0.284	29	2"410	0.291	14	1"110	0.299	10	0"812	0.470
Minkowsky3	21	1"687	0.309	13	1"687	0.309	72	5"316	0.293	37	2"963	0.264
Mahalanobis	35	2"687	0.567	15	1"172	0.547	17	1"812	0.567	14	1"189	0.557

Table 3 Results of classification for social spiders representation of data reduction dimension (CHI2) after encoding TFC

	N-grams character									1-gram word		
	2			3			4			1		
Distance	# Class	Learning time(S)	F-measure	# Class	Learning time(S)	F-measure	# Class	Learning time(S)	F-measure	# Class	Learning time(S)	F-measure
Jaccard	33	2"271	0.294	21	1"529	0.309	36	2"952	0.299	23	1"781	0.250
Cosine	38	2"979	0.264	27	2"359	0.243	101	6"793	0.215	99	6"765	0.212
Euclidian	10	2"812	0.275	14	1"110	0.299	20	3"292	0.217	19	1"497	0.323
Manhattan	20	1"516	0.255	57	4"514	0.228	20	1"516	0.273	20	1"516	0.232
Minkowsky3	13	1"687	0.299	11	1"812	0.279	32	3"257	0.302	31	3"048	0.280
Mahalanobis	16	1"184	0.657	16	1"184	0.543	25	1"983	0.652	30	3"025	0.672

4.4.4 Interpretation

We experienced 1000 documents of Reuters 21578 corpus with multiple representations of data, with and without dimension reduction before and after encoding TFC and several distances to try to identify sensitive parameters on the best classification results that validated the F-measure is a function of recall and precision. All tests on the parameters of data representation have been made to avoid misjudging our new approach based on a biomimetic approach in this case the social spiders because a good classification of data requires a good representation of the data.

The best classification results were obtained by the Mahalanobis distance as shown in tables 1, 2 and 3 (highlighted in yellow) validated by the biggest f-measure.

As regards the methods of data representation the best results were obtained by the 2-gram characters and the 1-gram word.

In terms of learning time, the method is promising because social spiders to 1000 to classify documents that time is between 1 second for the Mahalanobis distance and 6 seconds for the Jaccard index.

It should be noted that in the implementation of our approach we have used only artificial spider in a 2D environment with a random displacement. In other words, communication between agents is neglected. The system thus produced is seen as a deterministic system the likelihood of travel or weaving are omitted. So the execution of our application for the same parameters and the same data give the same results.

4.4.5 Comparison with Cellular Automata

In terms of recall and precision the best result is obtained by cellular automata for the cosine distance for a good result against clustering is obtained by social spiders to the Mahalanobis distance.

Table 4 Comparison of cellular automata and social spiders for text clustering

		Cellular Automata			Social Spiders	
Distance	# Class	Learning time(S)	F-measure	# Class	Learning time(S)	F-measure
Cosine	47	0.44	0.93	20	1.516	0.440
Euclidian	36	0.38	0.45	11	0.891	0.347
Manhattan	43	0.78	0.40	46	3.331	0.354
Minkowsky3				32	3.087	0.380
Mahalanobis				30	3.025	0.672

In terms of learning time, cellular automata are faster than the social spiders as well as the latter have a significant learning curve compared to other biomimetic methods.

4.4.6 Visual Appearance of Social Spiders

One of the strengths of our approach based on social spiders is the visual classes as shown in Figures 2,3,4 and 5.

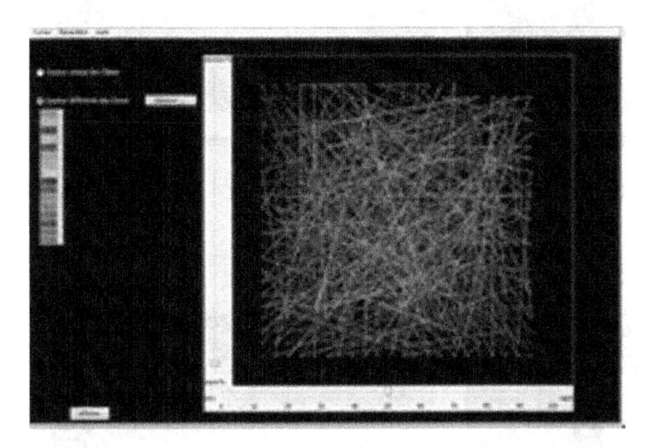

Fig. 2 Spiderweb for clustering of text represented by the 3-gram character and Mahalanobis distance.

It is clearly seen in all the graphs above the different colors that represent the classes (the son of woven spider) and the stakes represented by white dots that represent the documents. Stakes (documents) are visible in the figures zoomed.

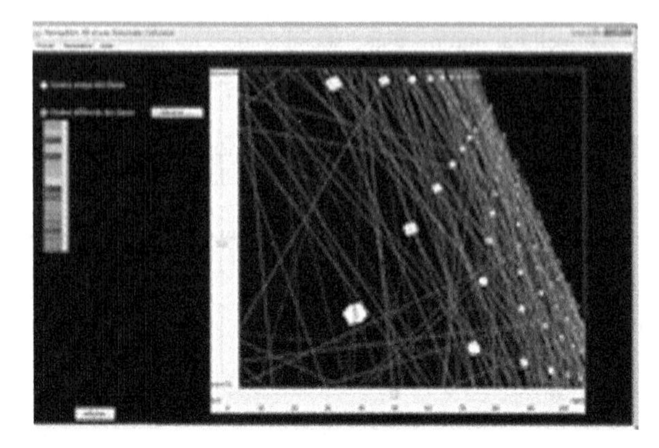

Fig. 3 Spiderweb for clustering of text represented by the 3-gram character and the Mahalanobis distance zoom.

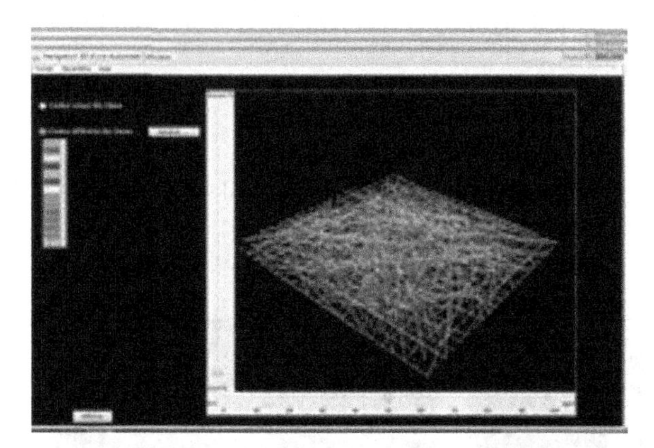

Fig. 4 Spider Web for clustering of text represented by the 4-gram character and Mahalanobis distance.

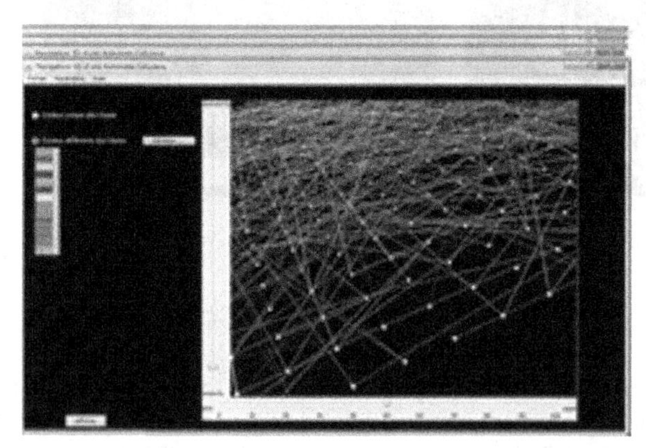

Fig. 5 Spiderweb for clustering of text represented by the 4-gram character and the Mahalanobis distance zoom.

5 Conclusion and Perspetives

In this paper the approach of social spiders is proposed as a solution to the problem of unsupervised classification (clustering) of textual records.

Movement and the weaving of the son of a stake to the next of artificial spider are changing the 2D grid of pegs formed into groups of similar documents whose path is none other than the cluster and the stakes that it contains no documents other than the clusters.

The experimental results are positive and confirms the idea of testing this new method biomimetic. The application is designed to visually navigate well (read the documents, Zoom, Rotate, ...) and do a reading class in the web woven by the spider.

The analysis and validation of classification results were based on evaluation criteria based on the notion of recall and precision in this case the f-measure.

Given the results, our approach based on a biomimetic approach (social spiders) can help solve one of the problems of data mining and text viewing.

In perspective we will try to develop this approach using a set of social spiders communicating with each other to form a multi-agent system capable of increasing the quality of the solution of clustering in other words use the principle of stigmergy. We will also attempt to explore other methods biomimetic, because nature has not yet revealed all the secrets to solve combinatorial problems still existing in the field of data mining.

References

1. Chen, H., Martinez, J., Ng, T.D., Schatz, B.R.: A Concept Space Approach to Addressing the Vocabulary Problem in Scientific Information Retrieval: An Experiment on the Worm Community System. Journal of the American Society for Information Science 48(1), 17–31 (1997)
2. Tan, A.-H.: Text mining: The state of the art and the challenges. In: Workshop on Knowledge Discovery From Advanced Databases, PAKDD 1999, Beijing, China, pp. 65–70 (1999)
3. Feldman, R., Dagan, I.: Knowledge discovery in textual databases (KDT). In: International Conference on Knowledge Discovery, Montreal, Canada, pp. 112–117 (1995)
4. Jain, A.K., Murty, M.N., Flynn, P.J.: Data clustering: A review. ACM Computing Surveys 31(3), 264–323 (1999)
5. Willet, P.: Recent trends in hierarchic document clustering: A critical review. Information Processing & Management 24, 577–597 (1988)
6. Rijsbergen, C.V.: Information retrieval, 2nd edn. Butterworths, London (1979)
7. Buhmann, J.: Data clustering and learning. In: Arbib, M. (ed.) The Handbook of Brain Theory and Neural Networks, pp. 308–312. The MIT Press, Cambridge (2003)
8. Hamou, R.M., Lehireche, A., Lokbani, A.C., Rahmani, M.: Representation of textual documents by the approach wordnet and n-grams for the unsupervised classifcation (clustering) with 2D cellular automata:a comparative study. Journal of Computer and Information Science 3(3), 240–255 (2010) ISSN 1913-8989, E-ISSN 1913-8997
9. Hamou, R.M., Lehireche, A., Lokbani, A.C., Rahmani, M.: Text Clustering Based on the N-Grams by Bio Inspired Method (Immune Systems). International Refereed Research Journal Researchers Worls 1(1) (2010) ISSN 2229-4686

10. Hamou, R.M., Lehireche, A., Lokbani, A.C., Rahmani, M.: Text Clustering by 2D Cellular Automata Based on the N-Grams. In: 1st International Symposiums on Cryptography, Network Security, Data Mining and Knowledge Discovery, E-Commerce and its Applications, October 22-24. Proceedinds IEEE Publishers, Qinhuangdao (2010)
11. Beni, G., Wang, U.: Swarm intelligence in cellular robotic systems. In: NATO Advanced Workshop on Robots and Biological Systems, Il Ciocco, Tuscany, Italy (1989)
12. Dorigo, M., Stützle, T.: Ant Colony Optimization. MIT Press, Cambridge (2004)
13. Clerc, M., Kennedy, J.: The particle swarm-explosion, stability and convergence in a multidimensional complex space. IEEE Transactions on Evolutionary Computation 6(1), 58–73 (2002)
14. Epstein, J.M., Axtell, R.: Growing Artificial Societies. MIT Press, Boston (1996)
15. Drogoul, A., Ferber, J.: Multi-agent Simulation as a Tool for Modeling Societies: Application to Social Differentiation in Ant Colonies. In: Castelfranchi, C., Werner, E. (eds.) MAAMAW 1992. LNCS, vol. 830, pp. 3–23. Springer, Heidelberg (1994)

Investigating Software Reliability Requirements in Software Projects

Marwa Abd Elghany, Anthony White, Elli Georgiadou, Nermine Khalifa, and Mona Abd Elghany

Abstract. Over the years, software has become ubiquitous in business processes from payroll systems to airline reservation systems. Software plays a vital role in facilitating business processes. Given the importance of these software systems, managing their quality is essential to the success of the business processes they support. Because quality attributes are important predictors of software quality as to provide a better understanding of the related features acquired for each software quality attribute, those features can be manipulated to improve the quality of a software project and determine the desired functional requirements necessary to satisfy the associated business processes. Accordingly, customer needs should be

Marwa Abd Elghany
Lecturer, Department of Business Information Systems, College of Management & Technology, Arab Academy for Science & Technology, Alexandria, Egypt
e-mail: marwam@aast.edu

Anthony White
Professor of Systems Engineering, School of Engineering and Information Science, Middlesex University, London, United Kingdom
e-mail: a.white@mdx.ac.uk

Elli Georgiadou
Principal Lecturer, School of Engineering and Information Science, Middlesex University, London, United Kingdom
e-mail: e.georgiadou@mdx.ac.uk

Nermine Khalifa
Associate Professor, Department of Business Information Systems,
College of Management & Technology, Arab Academy for Science & Technology, Alexandria, Egypt.
e-mail: nerminek@gmail.com

Mona Abd Elghany
Associate Professor, Department of Productivity and Quality Management,
College of Management & Technology, Arab Academy for Science & Technology, Alexandria, Egypt
e-mail: mabdelghany2000@gmail.com

R. Lee (Ed.): Software Engineering Research, Management and Appl. 2012, SCI 430, pp. 31–47.
springerlink.com

completely elicited in developing the software application then traced and referred back to throughout the software development process during the early requirements analysis phase. The proposed questionnaire empowers software developers to capture the functional reliability requirements and specify reliability related features for a software system.

Keywords: Software Quality, Customer Satisfaction, Measurement Concept, Reliability Attribute, Reliability Requirements.

1 Introduction

Quality is a complex concept because it means different things to different people; it is a highly context-dependent. Just as there is no way to satisfy everyone's needs, there is no universal definition of quality either. Reference [1] emphasizes that quality is an elusive target. Hence, there exists no single simplified measure of software quality that is acceptable to everyone. In order to assess the quality of a software in an organization, the quality aspects in field environment must be addressed, and then measured. Defining quality in a measurable way would make it easier for other people to understand the notified quality and their related business goals. And at the end, highly reliable software is determined by a mature process in a good successful business.

Over the last twenty years, there has been an increasing emphasis on quality in developing software ([2], [3], [4] & [5]). The quality of a software system is widely accepted as its conformance to customer requirements [6]. The interest in quality is heightened as more system failures are attributed to issues in software quality that often lead to higher maintenance costs, longer cycle times, customer dissatisfaction, lower profits, and loss of market share ([7], [8], & [6]).

Although the importance of quality is acknowledged, managing quality efforts remains a major challenge in software development. Despite the fact that most software developing firms collect quality performance measures such as customer satisfaction, but no operational measures are available for the quality attributes of software projects.

Reference [9] demonstrates that software quality has shown to affect user satisfaction, and therefore user satisfaction is often used as a surrogate for software quality. Reference [6] emphasizes the need to achieve total customer satisfaction through studying customer wants and needs, gathering customer requirements, and measuring customer satisfaction. Software quality aspirations are more likely to be achieved with a greater emphasis on customer satisfaction [10]. In addition, techniques such as quality function deployment (QFD) are heralded as a vehicle to help software developers hear the "voice of the customer" [3].

In an attempt to elicit the software reliability functional requirements needed to satisfy the customer requirements of the desired business processes, the authors depicts a questionnaire that provides a solid foundation for further empirical evaluation of the related features of the software reliability product. The following sections illustrate consecutively the adoption of knowledge elicitation into well specified software requirements then focusing on software reliability and its influencing factors as well as the different reliability views and concepts discussed in literature to end up with the description of the proposed questionnaire and briefing commentaries.

2 Knowledge Elicitation

The discipline of software metrics entails identifying various attributes that need to be measured and determining how to measure them in developing quality software [8]. Quality metrics must be utilized and tightly coupled with the software development process. Identifying the applicable body of knowledge required is the first step in equipping software engineers with the essential skill set. The transition from defining the "why" business objectives to defining the "what" business or functional requirements is the most challenging phase.

A business analyst should elicit requirements from different stakeholders to discover their concerns and needs. Short-change time spent on software requirements and listen to customer needs is proving to be very useful in order not to set project to failure. Most of the existing models adopted for software development process use the result of design, implementation and test phases; whereas the assessment of the software desired characteristics in the early phase of software development process would better support risk management and cost estimation associated with software projects.

Measurement is the key to achieving high quality software. The software engineer would apply the body of knowledge elicited to improve the quality of software throughout the development life cycle. In addition, the body of knowledge may be used as guidelines for practitioners, licensing of software professionals, and for training in software quality measurement. Lack of knowledge could result in significant costs to the supplier in terms of unsatisfied customers, loss of market share, rework caused by rejected and returned systems, and the costs to customers of faulty systems failed to meet their mission goals. The interpretation of the desirable properties of a software product in quantitative terms is an important part of the software requirements engineering in the modern world. Because once the product is released, it is no longer in a controlled test situation but instead in-practice with different users.

The concept of measurement scales is vital in software data assessments [11]. Software measurement theory addresses the issue of whether the proposed metrics are valid with respect to the attributes of the software entitled to measure. These evaluations are based on the properties of the measurement scales. Various development artifacts, such as requirements, design, and code documents have disclosed the majority of software faults prior to testing and enhanced the ability to make meaningful assessments and predictions of product reliability quality. As a result, the principle of software measurement is the use of substitute measures that could be available early during evaluation process and hypothesized to be representative of the measures of real interest.

Concurrently, the system goals of the organization should be analyzed in order to understand how to specify the system requirements. This analysis could consist of interviewing key personnel in the organization and examining documentation that addresses these goals. As defined by [12] "A quality requirement is a requirement that a software attribute be present in software to satisfy a contract, standard, specification, or other formally imposed document". For example, a goal in a safety critical system could be translated into the nonexistence of neither fault nor failure that would jeopardize the mission and cause loss of life. Hence the application domain has implications for system requirements, and the context in which the software will be functioning in must be well understood. In case of reliability attribute, the influence of different hardware configurations; standalone, network, fault-tolerant, etc. on software reliability requirements should be considered. For example, a network would have higher reliability and availability requirements than a standalone because unreliable operation in a network could affect thousands of users simultaneously, whereas it would affect only a single user at a time in a standalone operation. Thus, software engineers must be aware of the quality of service needs of the user organization.

3 Software Reliability Definition

Software reliability could be a difficult concept to grasp. A software product will fail under certain conditions, with certain inputs, and given the same inputs and conditions will fail every time until the cause of the failure is corrected. Accordingly, the reliability of a software product is more about the discovery of faults resulting from various inputs with the system in various states. In other words, an estimate of the actual performance of the software system in a future production environment is obtained by the measurement of the number of failures per hour or per transaction in a system test environment. Reaching an optimal performance free of failure for at least a given period of time or an error-free session for a certain timeline can clearly define the related features of a software reliability product. Reliability attribute targets a minimized failure rate in order to enable full functionality of software application and achieve package's deliverables under the specified condition of use.

Reference [13] defined reliability as the “probability of failure-free software operation for a specified period of time in a specified environment”. Early software reliability assessment is achieved by the reliability measurement at the requirement phase. The number of faults found during the requirements analysis and software design have to be normalized by the code size estimates and expressed in terms of fault densities [14].

Therefore the reliability of a system is usually defined by its ability to provide a failure-free operation and simultaneously for measuring the reliability of a product, accurate and complete failure data must be collected as well as the population size, declaiming this data, is needed. Often the failure data is obtained through a Product Service Organization where users can report failures when they encounter them, and population data is obtained from the sales figures ([15], [16], & [17]). This implies that reliability of a product is the same for all users and that most failures are reported besides the user base is known. However, these assumptions are infeasible for a mass-market product as there are different users' groups with widely varying operational profiles, having different rating preferences to the reported data failure. The weight assigned to the failure type changes from product to product. For example, for some products a user-interface failure is important while for real-time applications operational performance failure is far more important.

Providing a version of software with minimal failure rate is the common description of reliability attribute. Reference [18] estimated the failure rate of a software product in this equation form ($\lambda = F / (N * T)$) where F is the summation of failure cases for several installation N within limited timescale (generally assumed to be 24 hours). This approach for measuring failure rates has been widely used ([15], & [17]). However, it has the same limitations illustrated in the previous paragraph. In addition, the usage time of software is not 24-hours a day and the number of software installations is not well triggered as claimed.

Despite of the "perfect" design that might be allocated to software packages, a regular maintenance and upgrade should be accomplished for any software package. Reference [19] assured that there are no means to stop an "ultimate failure" of any engineered object; a man-made entity will fail in due course. Reference [20] defined a failure as an incident that causes an undesired status for a service of a product when the intended purposes are not performed safely, and cost-effectively. Such accidents or failures may occur due to some unexpected technical or mis-operational causes. Mean Time Between Failures (MTBF) is used as a measure for a degree of reliability of a certain product. Greater reliability degree means large span between failures, and vice versa [21]. As can be seen, software reliability is associated mainly with hardware frameworks and its practice is obviously related to tangible resources and failures. Thus it might be desirable for hardware reliability theory to be applied to software.

4 Failure Types Classification

Different types of failures must be distinguished and reflected in accordance to user perception and several users' scenarios should be issued to signify the reality. With more focus on user requirements and their associative weights for each failure type to represent its relative importance to the product application type, measurement in this form, can help better in determining the product parts needed to be repaired as well as other issues to be strengthened that will indeed enhance the product reliability.

Though many classifications have been presented in the literature [22] and the IEEE standard [23], unrealistic assumptions limited its boundaries. A taxonomy for failures is needed from the users' perspective. For example, reference [24] categories types of failures into omission, timing, and response failure. The user-oriented perspective was not considered in such scheme. Reference [25] developed multi-tier taxonomy that defines three main categories: unplanned events, planned events, and configuration failures. Unplanned events are traditional failures like crash, hang, incorrect or no output, which are caused by software bugs. While the planned ones are those being handled in a pre-defined manner like updates or configuration changes that require a restart to make file registry. Configuration failures usually occur due to application/ system incompatibility error or installation/ setting failures. In many systems, configuration failures account for a large percentage of failures [26]. It has been acknowledged that the user experience and familiarity with the system minimizes the probability of causing failure.

Earlier, product service organization repositories were used to record the different types of failures, reported by different users groups ([15], [16], [27], & [17]. With the consideration, that not all users report every single failure as they may resolve them by themselves and that failure which is not included, in case of the existence of multilevel PSOs, will be forwarded to the product vendor to be resolved. Therefore following such methods for obtain a full pattern analysis of product reliability will be infeasible. Instead, event logging mechanism is being used to record specific events. These events assess user interactions for the detection of the usage time, program state and exit status in order to monitor failure data. The product vendor would filter such events along with the other operating systems recorded events like reboots and crashes to segment the failure case for maintenance. Simultaneously, the availability and reliability of product can be easily measured ([28], [29], [30], & [31]).

Moreover, it is well recognized that the failure rate of software often depends on the capacity of the hardware ([32], & [33]). Consequently; the system configuration must be considered and characterized for the improvement of the product reliability.

Reference [25] recognized several types of event failures. These failures might differ from total software failures, like crashes and hangs, to asserts and notification alerts stating certain functionality decline, as explained in the next coming bullets:

- Application termination – records application closure like normal exits, crash exits, hangs, and user enforced exits. Such events are being tracked and documented either in time of occurrence or with the startup of predecessor session.
- Assert failures – records statement-related failures embedded in the program source line of codes. Those assert statements are called ship asserts as they are shipped to the customers in the application code.
- Alerts – notifying users in case of resource unavailability or failure related to an action handling (e.g. file does not exist, network not available, file writing fails, etc.). Information of such failures is being recorded for reporting service.

In addition to the recording of failure events, information about used resources is also being documented as well. This information will indicate the used platform, hardware configuration and software version. Referring to cases that were triggered through event logging mechanism along with the information collected of planned events, an entire picture about failure cases and their causes will be provided. Other attributes could be derived using such mechanism like the session timeline, count of user sessions, user inactive time per session and actual usage to indicate the performance ratio of individual application.

Crashes and hangs failures are most commonly highly considered by system users and managerial levels. Different parameters related to application in which the failures occur, such as the displayed notification alerts and the involved users in each session, are being captured by logging mechanism. Failure cases attached to the configurations assigned and events occurred compose packages that might enable developers to figure out the reasons behind such malfunctioning and fix such errors. Shutdown Event Tracker is being used as well to keep track of restores and downtimes [31]. Giving different weights for different failure types is applied to assess the software reliability. An accumulative way of weighting reliability features is embraced for product evaluation and individual analysis of usage pattern. Comparative studies between various products and same products but different versions will be possible in accordance with such cumulative weights measurement method.

The overall user satisfaction is affected with the technical software reliability as well as other factors; like ease of installation, product learning curve, management flexibility and interoperability or the application ability to integrate with other applications. Reliability measurements help in the assessments of such factors. For example, the simplicity of installation is commonly measured with the timeline needed to reach a stable state while the number of reboots taken to perform any action is employed to determine the flexibility of software management. The next coming section will discuss such factors in more detail.

5 Factors Affecting Software Reliability

The assessment of system functionality means measuring its probability of working properly under a predefined condition, i.e. reliability could be considered a good performance measures for a system. There exist various models that are measuring this probability density using statistical methods [34]. Such approaches required the collection of quite a huge amount of data to provide reliable estimates. Recently, most of the developed models focus on fault density and test reporting as well as debugging processes [35]. Markov model, Halstead's Software Metric, McCabe's Cyclamate Complexity Metric, Poisson probability theory, are the most common approaches that extended the statistical models to identify the dynamics of the software product reliability [36].

Reference [37] developed a mathematical function called system ability, which considers the uncertainty of the operational environments in the system reliability prediction. Other factors like the Difficulty of Programming, the Level of Programming Technologies and the Percentage of Reused Code must be considered as well. Reference [38] identified additional aspects including availability, accessibility, and performance that should be present in the software reliability evaluation and deduced into the following metrics such as: maturity metrics, fault tolerance metrics, recoverability metrics and compliance metrics.

Reference [35] referred to the following indicators in the description of reliability. System Mean-Time-To-Failure (SMTTF) is the most common used measure to assess the software reliability. Recoverability is another indicator that refers to the ability to reinstate a system in accordance with predefined procedures and resources. How to repair a fault and return back to the ready-state of system formulizes recoverability concept. Time needed for this restoring operation as well as the necessary skill with the support equipment and documentation should be determined for such part removal or replacement then reestablishment of the product under analysis. So the reliability attribute focuses on system functionality while availability is more concerned with system recovery.

Consequently, specialists should be concerned with the data analysis of the operating times, failure rates, and recovery times as well as their attention to sparing and maintenance hardware support. Design reviews and compliance documents should focus on the software-related sustainment issues of computer systems which are defined in terms of availability or dependability related reliability attribute as such systems involve highly complex software [39]. It could take a definite mission time as short as minutes for a tactical missile, or as long as years for a space vehicle [40]. System monitoring, runtime software related diagnostics, restarts, and reloads are essential to achieve high availability.

Software specifications are the authoritative statements of software requirements. Requirements for reliability are stated either as a probability of successful execution over a defined time interval i.e., mean time to failure (MTTF, for a non-repairable system), mean time between failures (MTBF, for a repairable system), mean mission duration, or life design. In case of software-intensive systems with mission durations of hours to years, there are multiple modes and states. Whereas for small systems there is one main execution loop, with mission durations on the order of minutes, consequently, most software failures result in a system failure. For example, in a satellite, a software failure might result in a loss of an on-board instrument, communication channel, or even more severe cases that would be recoverable if the satellite safing mechanism is effective. Therefore, it is necessary to specify multiple failure conditions and the appropriate time interval for each. In other words, there must be a differentiation between a "critical failure" and "downing event", or other designations for a loss or a degradation of service but not necessarily cause its removal from operation. Downing events could result from preventive maintenance, training, maintenance and supply response, administrative delays, and actual on-equipment repair.

Availability, or in other words, dependability requirements are difficult to specify due to the following reasons [41]:

- the presence of multiple states for the system and their influence on the mission;
- the need to distinguish between the loss of a service at a single operational position and the central service lost at all operational positions;
- the multiple types of operating time definitions; what modes and states are involved in operating hours: if no services are executing – is it just adequate that the hardware be functional?
- the difficulty of differentiating between operator induced failures, or in other words, failures caused due to error input and those induced by the system hardware or software itself.

For long duration missions, availability metrics are determined by the severity of the failures, in question, that could be measured by the frequency and duration of recoverable events. Reliability standards should include allocations for software recovery times, with the consideration that shortening recovery times can often be more easily achieved than increasing MTTFs. Moreover, hardware failures should be also included along with software failures, and both of their effects are recorded whether or not they are reproducible.

Nevertheless, much of the software reliability and related performance parameters in current generation of software- intensive systems are predetermined and measurements can be made in a representative environment. Examples include the underlying operating system, messaging "middleware" layers, network monitoring and control software, and database management systems. Methods for measurement of developed software in the context of an operational environment are also well established in ref. [42] and [43].

Reference [41] suggested the collection of the following to ensure the effective use of failure data and its completeness:

- all instances of failures (whether or not reproducible);
- operating time of all software tasks and processes;
- recovery time for each recoverable failure (and the affected components);
- for each failure, whether detection and recovery occurred without manual intervention;
- in redundant systems, proportion of failures that were common mode or common cause.

Automated logging capabilities of modern multitasking operating systems enable the complete collection of such data for subsequent analysis.

6 Software Quality Models Views towards Reliability Requirements

In the software requirements engineering, a software engineer should specify and describe the non-functional requirements describing the behavior of the system functions into functional requirements defining the required system functions to be executed then develop, test, and configure the final deliverables to system users.

ISO 19761 [44] proposes a generic model of software functional requirement that clarifies the boundary between hardware and software. Software is bounded by persistent storage hardware like a hard disk, RAM or ROM memory. Its functionality is embedded within the flows of data groups that can be characterized by four types of data movements. The exchange of data with users across a boundary (i.e. at the front-end) is allowed via Entries and Exits movements while at the back-end interface, Reads and Writes movements allow the exchange of data with the persistent storage hardware. Simultaneously, users could be considered the I/O hardware that interacts directly with the software. COSMIC – ISO 19761 [44] is aimed to measure the size of software based on identifiable functional requirement.

Various ECSS (European Cooperation on Space Standardization) standards for the aerospace industry ([45], [46], [47], [48] & [49]) express the related elements of reliability in different system views. Reliability is considered as the acceptable probability of system failure which is based on the equipment reliability and availability specifications in order to check the following:

- Failure modes and effects,
- Failure tolerance, failure detection and recovery,
- Statistical failure data to support predictions and risk assessment,
- Capability of the hardware to operate with software or to be operated by a human being in accordance with the specifications,
- Demonstration of critical items reliability, and
- Justification of data bases used for theoretical demonstrations.

The table below lists the concepts used in the ECSS standards that describe system-related reliability requirements.

Table 1 Reliability view and vocabulary in ECSS [50]

Key views	Concepts and Vocabulary
Acceptable probability of system failure	▪ Component failure ▪ Redundancy feature ▪ Data parameter ▪ Reliability methods, operations and mechanism ▪ Failure tolerance ▪ FMEA and FMECA ▪ Failure detection ▪ Failure isolation ▪ Failure recovery ▪ Failure data

IEEE-830 [51] states the reliability requirements as the factors that establish the required reliability of the software system at time of delivery.

IEEE-1220 [52] articulates the reliability requirement as the analysis of system effectiveness for each operational scenario.

ISO 9126 [53] identifies reliability as a 'quality characteristic', which is decomposed into sub-characteristics that can be interpreted into derived measures.

Table 2 presents the related concepts and vocabulary on software reliability, such as maturity, fault tolerance and recoverability that were introduced in ISO 9126.

Table 2 Reliability view & vocabulary in ISO 9126 [50]

Key views	Concepts and Vocabulary
The capability of the software product to maintain a specified level of performance when used under specified conditions	▪ Maintain ▪ Fault tolerance ▪ Recoverability ▪ Fault Density ▪ Failure Resolution ▪ Incorrect Operation ▪ Availability ▪ Breakdown Time ▪ Recovery Time ▪ Fault Removal ▪ Failure Avoidance ▪ Restart ability ▪ Restorability

In addition, ISO 24765 [49] for the systems and software engineering vocabulary refers to the reliability as the probability that software will not cause the failure of a system for a specified time under specified conditions and conceptualizes two functions; one to identify error to input and the other to identify error to output.

Table 3 aggregates the common system reliability requirements that are present in the ECSS standard with the reliability-related concepts found in ISO 9126.

Table 3 Reliability in ECSS, IEEE & ISO 9126 [50]

Functions to address system reliability requirements
Function to identify error to handle input
Function to identify error to produce output
Function to identify error to produce correct output
Function to identify fault prevention
Function to identify fault detection
Function to identify fault removal
Function to identify failure operation
Function to identify failure mechanism

Reference [50] derived from the above concepts (system prediction tolerance, system maturity, system fault tolerance, and system recoverability) several reliability functions to be specified and interpreted into corresponding entities to be measured. These functions are categorized into external and internal reliability functions that may be allocated to software (as shown in Table 4) where:

- External reliability refers to the reliability prediction for faults, failures and errors that could be found in the system.
- Internal reliability refers to the reliability assessments for faults, failures and errors found in the system.

Table 4 Reliability functions that may be allocated to software [50]

Types of reliability functions	Reliability Functions
External Reliability	Error tolerance Fault tolerance Failure tolerance
Internal Reliability	Error to handle input Error to produce output Error to produce correct output Fault prevention Fault detection Fault removal Failure operation Failure mechanism

This categorization for reliability requirements and their allocation to software functions implementing such requirements signified the main headlines that were adopted in developing the proposed questionnaire for specifying and measuring software requirements needed to address the system reliability.

7 Questionnaire Design

Requirements that are shown in the following questionnaire presents some of the important functions that would be performed by the software engineer in executing a life cycle reliability management plan. These functional requirements were identified by information systems professionals who had an expertise in practice for software development or in other words, had been involved in different aspects of software development. Those contributors concluded that it is essential to address these requirements if a software developer is to be successful in producing high reliability software.

The respondents to the questionnaire were requested to rate each item in the instrument on a five-point Likert scale ranging from "strongly disagree" to "strongly agree" based upon their perception. Considering the unsolicited nature of the field survey, the length of the questionnaire and the sheer number of research surveys targeted at IS professionals (including requirements' engineers, system designers, software developers, system operational support, software maintenance specialists, testers, etc.) whom currently and previously were engaged in managing software development projects from the Information and Documentation Centre and also staff members of the Computing & Technology Faculty, in the Arab Academy for Science and Technology (AAST) due to their expertise in programming and analysis, the response rate was deemed satisfactory and quite comparable to other similar studies. The level of experience and knowledge of each participant was evaluated using a user profile subdivision. Also, all participants were unpaid volunteers who had professional interest in software requirements engineering. However, it should be noted that an effort has been made to provide a comprehensive description of the questionnaire protocol so that empirical evaluation can be in depth investigated and replicated in future work.

Here is the produced questionnaire presenting a number of statements related to the most functional reliability features. The participant is requested to indicate his/her response to the degree of importance of each statement to reliability requirements specification by ticking (✓) the suitable number: 5 =Very High, 4 =High, 3 =Fair, 2 =Low, 1 =Very Low. In case of no applicability of the statement please tick ✓ the Not Applicable (NA) column.

Table 5 Produced questionnaire [The Authors]

Question Element		Importance Degree				
First: Questions relating to monitoring failure type using the event logging mechanism	NA	1	2	3	4	5
1. Identifies error to handle input						
2. Detects error to produce output like hangs						
3. Encounters error to produce correct output like crash failures						
4. Registers application closure like normal exits, and user enforced exits						
5. Records statement-related failures embedded in the program source line of codes						
6. Notifies the users in case of resource unavailability or failure related to an action handling (e.g. file does not exist, network not available, file writing fails, etc.)						
7. States the used platform, the hardware configuration and the software version in which the failure occurred						
Second: Questions relating to measuring performance ratio of software application	NA	1	2	3	4	5
8. Gives the session timeline						
9. Counts the number of user sessions						
10. Records the user inactive time per session						
11. Scores the actual user usage						
Third: Questions relating to availability through runtime diagnosis and shutdown event tracker to trace downtimes and restores	NA	1	2	3	4	5

Table 5 (*continued*)

12. Reports time taken to remove or repair a fault for returning back the system to its ready-state						
13. Allocates the recovery times for the annotation of the number of faults						
14. Normalizes the number of faults by the code size estimates to be expressed in terms of fault densities						
15. Assesses configuration changes that require a restart to make file registry						
16. Identifies fault prevention that could result from preventive maintenance						
17. Recognizes fault prevention resulting from supply response						
18. Pinpoints fault prevention resulting from administrative delays						
Fourth: Questions relating to interoperability to integrate with other applications and management flexibility	NA	1	2	3	4	5
19. Detects error due to installation setting configuration						
20. Provides the timeline needed to reach a stable state						
21. Discloses the number of reboots taken to perform any action is employed						
Fifth: Questions relating to system ability of the operational environments	NA	1	2	3	4	5
22. Considers the Difficulty of Programming						
23. Indicates the Level of Programming Technologies						
24. Detects the Percentage of Reused Code						

8 Concluding Remarks

The appropriate definition of requirements for software-system reliability in specifications of programmatic requirements in contractual documentation can avoid most of the problems arising after development phase during the practical use of the software.

For the users of a system it is the reliability of the system as a whole that is meaningful but for analysts and testers it is important to separate the software requirements from the hardware requirements as there are some significant differences. There is no question that any piece of hardware will eventually wear out and fail so hardware product reliability could be stated in terms of time to failure or time between failures, with the assumption that pieces of hardware will continue to operate for various lengths of time before failure.

Nowadays, dealing with standard software packages that are running on similar configuration of servers by a defined number of users is an invalid assumption. There is a different variety in the operational platforms, users' preferences of settings and priorities that should be taken into consideration. Following the illustrated approach of weighting to end up with a tangible product reliability measure allows the convention of reliability measure into a patronized form that is consistent with the real environment. More research is required to verifying validity.

The questionnaire can be useful for both researchers and practitioners in grasping the critical related features of software reliability. Researchers can benefit from the instrument in terms to the quality of the software being built. It provides a road-map for improving software reliability as well. Specifically, a manager could administer the software project periodically, assess changes and consequently take the appropriate action to ensure continuous improvement in software quality.

References

[1] Kitchenham, B.: Measurement for Software Process Improvement. Blackwell Publishers (1996)

[2] Duggan, E.: Silver pellets for improving software quality. Information Resource Management 17(2), 60–95 (2004)

[3] Haag, S., Raja, M., Schkade, L.: Quality function deployment usage in software development. Communications of the ACM 39(1), 41–49 (1996)

[4] Harter, D., Slaughter, S.: Quality improvement and infrastructure activity costs in software development. Management Science 49(6), 784–796 (2003)

[5] Prajogo, D., Sohal, A.: The integration of TQM and technology/R&D management in determining quality and innovation performance. Omega 34(3), 296–312 (2006)

[6] Kan, S., Basili, V., Shapiro, L.: Software quality: An overview from the perspective of total quality management. IBM Systems Journal 33(1), 4–19 (1994)

[7] Arthur, J.: Improving software quality: An insider's guide to TQM. Wiley, New York (1993)

[8] Gopal, Mukhopadhyay, T., Krishnan, M.S.: The Role of Communication and Processes in Offshore Software Development. Communications of the ACM 45, 193–200 (2002)

[9] DeLone, W., McLean, E.: Information systems success: The quest for the dependent variable. Information Systems Research 3(1), 60–95 (1992)

[10] Lin, W., Shao, B.: The relationship between user participation and system success. Information & Management 37(6), 283–295 (2000)

[11] Zuse, H.: A Framework of Software Measurement, Walter de Gruyter, Berlin (1998)

[12] IEEE Standard for a Software Quality Metrics Methodology, Revision, IEEE Std 1061-1998 (December 31, 1998)

[13] Isazadeh, D., Lamb, A., Shepard, T.: Behavioral views for software requirements engineering. Requirements Engineering Journal 4(1), 19–37 (1999)
[14] McCall, J.: Rome Laboratory (RL), Methodology for software reliability prediction and assessment. Technical Report RL-TR, vol. 1, 2, pp. 92–52 (1992)
[15] Chillarege, R., Biyani, S., Rosenthal, J.: Measurement of failure rate in widely distributed software. In: Proc. 25th Fault Tolerant Computing Symposium, FTCS- 25, pp. 424–433 (1995)
[16] Gray: A census of Tandem system availability between 1985 and 1990. IEEE Transactions on Reliability 39(4), 409–418 (1990)
[17] Wood, P.: Software Reliability from the Customer View. IEEE Computer, pp. 37–42 (August 2003)
[18] Trivedi, S.: Probability and Statistics with Reliability. In: Queuing and Computer Science Applications, 2nd edn., John Wiley and Sons (2002)
[19] Blischke, W.R., Murthy, D.N.P.: Reliability, Modeling, Prediction, and Optimization, p. 3. John Wiley and Sons Inc., New York (2000)
[20] Isaic-Maniu, A.: Reliability and Its Quantitative Measures. Informatica Economică 14(4) (2010)
[21] Witherell, C.E.: Mechanical Failure Avoidance, p. 11. Mc. Graw-Hill Book Co., New York (1994)
[22] Chillarege, R., et al.: Orthogonal defect classification – A concept for in-process measurements. IEEE Trans. on Software Engineering 18(11), 943–956 (1992)
[23] IEEE, IEEE Guide to Classification for Software Anomalies, IEEE Standard 1044.1 (1995)
[24] Cristian: Understanding fault-tolerant in distributed systems. Communications of the ACM 34(2), 56–78 (1991)
[25] Jalote, P., Murphy, B., Garzia, M.R., Errez, B.: Measuring Reliability of Software Products. In: ISSRE 2004 Conference, Saint-Malo, Bretagne, France (2004)
[26] Chillarege, R.: What is software failure. IEEE Transactions on Reliability 45(3), 354–355 (1996)
[27] Kan, S., Manlove, D., Gintowt, B.: Measuring system availability – field performance and in-process metrics. In: Supplementary Proceedings of ISSRE 2003, pp. 189–199 (2003)
[28] Garzia, R.: Assessing the reliability of windows servers. In: Proc. Conference on Dependable Systems and Networks, DSN, San Francisco (2003)
[29] Iyer, R.K., Lee, I.: Measurement-based analysis of software reliability. In: Lyu, M.R. (ed.) Software Reliability Engineering, pp. 303–358. McGraw Hill and IEEE Computer Society Press (1996)
[30] Murphy, Gent, T.: Measuring system and software reliability using an automated data collection process. Quality and Reliability Engineering International (1995)
[31] Murphy, Levidow, B.: Windows 2000 dependability. In: Proc. IEEE DSN (June 2000)
[32] Iyer, R.K., Butner, S.E., McCluskey, E.J.: A statistical failure/load relationship: results of a multi-computer study. IEEE Trans. on Computers C-31, 697–706 (1982)
[33] Iyer, R.K., Velardi, P.: Hardware-related software errors: measurement and analysis. IEEE Tran. on Software Eng. SE-11(2), 223–231 (1985)
[34] Zhang, X., Pham, H.: An analysis of factors affecting software reliability. Journal of Systems and Software 50, 43–56 (2000)
[35] Pham: System software reliability. Springer (2006)

[36] Cotfas, L.A., Diosteanu, A.: Software Reliability in Semantic WebService Composition Applications. Informatica Economică 14(4) (2010)
[37] Pham: A new generalized system ability model. International Journal of Performability Engineering 1(2), 145–155 (2005)
[38] Canfora, Penta, M.D., Esposito, R.: An approach for QoS-aware service composition based on genetic algorithms. In: Genetic and Evolutionary Computation Conference, pp. 1069–1075 (2005)
[39] U.S. Government Accountability Office, Concurrency in Development and Production of F-22 Aircraft Should Be Reduced, GAO/NSIAD-95-59 (April 1995)
[40] U.S. Government Accountability Office, Future Combat Systems Challenges and Prospects for Success. Statement of Paul L. Francis Before the Subcommittee on Tactical Air and Land Forces, Committee on Armed Services, House of Representatives, GAO-05-428T (March 16, 2005)
[41] Hecht, M., Owens, K., Tagami, J.: Reliability-Related Requirements in Software-Intensive Systems, pp. 155–160. IEEE (2007)
[42] Lee, I., Tang, D., Iyer, R.K., Hsueh, M.-C.: Measurement-Based Evaluation of Operating System Fault Tolerance. IEEE Trans. Reliability 42(2), 238–249 (1993)
[43] Tang, Hecht, H.: An approach to measuring and assessing dependability for critical software systems. In: Proc. International Symp. on Software Reliability Engineering, November 2-5, pp. 192–202 (1997)
[44] ISO/IEC-19761, ISO 19761: Software Engineering- COSMICv3.0-A Functional Size Measurement Method, International Organization for Standardization, Geneva, Switzerland (2003)
[45] ECSS-E-40-Part-1B, Space Engineering: Software–Part 1 Principles and Requirement, European Cooperation for Space Standardization, The Netherlands (2003)
[46] ECSS-E-40-Part-2B, Space Engineering: Software – Part2 Document Requirements Definitions, European Cooperation for Space Standardization, The Netherlands (2005)
[47] ECSS-Q-80B, Space product assurance: Software product assurance, European Cooperation for Space Standardization, The Netherlands (2003)
[48] ECSS-ESA, Tailoring of ECSS, Software Engineering Standards for Ground Segments, Part C: Document Templates, ESA Board of Standardization and Control (BSSC) (2005)
[49] ECSS-E-ST-10C, Space engineering: System engineering general requirements, Requirements & Standards Division Noordwijk, The Netherlands (2008)
[50] Al-Sarayreh, K.T., Abran, A., Santillo, L.: Measurement of Software Requirements Derived from System Reliability Requirements. In: ECOOP 2010, Maribor, Slovenia, EU. ACM (2010)
[51] IEEE-Std-830, IEEE Recommended Practice for Software Requirements Specifications. IEEE (1993)
[52] IEEE-1220, IEEE Standard for Application and Management of the Systems Engineering Process, 1st edn. IEEE Computer Society (2007)
[53] ISO/IEC-9126, Software Engineering - Product Quality Model. International Organization for Standardization, Geneva, Switzerland (2004)

A Design Pattern Detection Approach Based on Semantics

Imène Issaoui, Nadia Bouassida, and Hanêne Ben-Abdallah

Abstract. Design patterns are good design solutions to recurring problems. Many works were interested in design patterns identification either for reverse engineering purposes, or for design improvement purposes. All existing approaches considered that a pattern is detected through both its structure and behavior, but no one considers the semantic aspect conveyed by the class and method names. In this paper, we propose a technique that exploits the semantic aspect to identify occurrences of a pattern in a design. In addition to the structural and behavioral analyses, the semantic analysis is very useful, specifically when there is a doubt between two design patterns having similar structures. By resolving a non deterministic identification, the identification precision increases.

Keywords: Design pattern detection, semantic aspect.

1 Introduction

Design patterns offer good solutions to recurrent design problems [1]; they improve the quality of software systems and facilitate system maintenance by helping to understand design and implementation issues. In order, to benefit from design patterns, it is useful to assist an inexperienced designer to improve his/her design by identifying design patterns that could better structure his/her design. Indeed, several researchers have proposed techniques to automatically recover design patterns. They perform either a static detection by considering the structural aspects of a pattern (e.g., [2] [3] [4] [5], or they combine a static and dynamic detection (e.g., [6] [7] [8] [9] [10] [11] by considering both structural and behavioral requirements. However, current propositions neglect the semantic aspect which can be expressed through the names of classes or methods in a design pattern.

Through our experience in pattern detection, we found that the detection of design patterns requires an analysis of the semantics in addition to the structure and

Imène Issaoui · Nadia Bouassida · Hanêne Ben-Abdallah
Mir@cl Laboratory, University of Sfax, Tunisia
e-mail: imene.issaoui@gmail.com, Nadia.Bouassida@isimsf.rnu.tn, Hanene.BenAbdallah@fsegs.rnu.tn

R. Lee (Ed.): Software Engineering Research, Management and Appl. 2012, SCI 430, pp. 49–63.
springerlink.com

behavior. In fact, the semantic aspect conveyed by the class and method names characterizes the pattern and it distinguishes one pattern from another. For example, the State and Strategy patterns have the same structure, but they can be set apart since they have different behavior and also since the class and operation names have different semantics and meanings. However, in existing works (e.g., [2] [3]), there is no distinction between the State and Strategy patterns. For similar reasons, Tsantalis et al. [2] do not differentiate between the Adapter and Command patterns.

To detect with no ambiguity a design pattern, one needs to identify first the structure or its variations, then the behavior of the pattern. In addition, one needs to validate the results of these two identification steps through the semantic analysis.

In this paper, we illustrate how our approach based on an XML document retrieval technique (presented in [12]) can be improved to identify design patterns without any confusion thanks to the semantic analysis. Being based on an XML retrieval technique, our approach identifies the similarities between the structure of the pattern and a given design. It has the advantage of detecting similarities between a group of classes in the application and the pattern while tolerating some variation of the pattern in the design. To do so, our approach relies on a context resemblance function [13] to compute the similarity potential between the application structure and behavior, and the pattern. The resemblance function is applied so as to account for both the structure and dynamic aspects of the pattern. In addition, it accommodates design variability with respect to the pattern structure without losing the pattern essence. The main objective of this paper is to enrich this approach by a semantic identification step.

The remainder of this paper is organized as follows: Section 2 summarizes related work; Section 3 overviews our design pattern detection approach [12] and shows how it can be improved by accounting for the semantic aspect. Section 4 details the semantic identification step. In section 5, our approach is illustrated through an example Finally, Section 6 concludes with future work.

2 Related Work

Several approaches were proposed in the literature to recover design pattern instances by performing a static analysis of structural aspects of the patterns (e.g [3] [5]). Other approaches [10] [6] [11] carried out the design pattern identification using either dynamic analysis or a combination of static and dynamic analysis.

The design pattern detection approach DeMIMA [3] is based on a multilayered approach able to identify design motifs. Note that, the authors used the term design motif to indicate the design pattern micro-architecture. The first layer of DeMIMA is devoted to the construction of models from source code, by using a language whose meta-model is inspired by UML. In the second layer, programming idioms are identified to define specific characteristics of classes or relationships between them. In the third layer, the language used to describe the source code is also used to represent design motifs, and micro-architectures similar to the specified design motifs are recovered from the model of the source code; this step uses explanation-based constraint programming and constraint relaxation. The main limitation of DeMIMA is that it focuses only on the structural aspect of the patterns, while neglecting the behavioral aspect.

Tsantalis et al. [2] consider that recognizing a pattern in a design is a graph matching problem. They propose a design pattern detection approach based on similarity scoring between graph vertices. The authors use a similarity scoring approach, inspired from the similarity scoring algorithm originally proposed by [14]. The graphs of the searched pattern and the examined design are encoded as matrices. These latter are used to compute a similarity matrix. The main limitation of this approach is that the algorithm can only calculate the similarity between two vertices, instead of two graphs, which can generate identification errors. In addition, another drawback of this approach is the convergence time which can be huge.

Dong et al. [4] use a template matching method to calculate the normalized cross correlation between the pattern matrix and the matrix representing a design segment. A normalized cross correlation shows the degree of similarity between the pattern and the design. This approach encodes eight design features, such as generalization, association and method invocations.

Belderrar et al. [5] propose an approach and a tool, named SGFinder that infers OO micro-architectures from class diagrams. Thus, they model a class diagram as a directed graph and define a micro-architecture as the connected subgraph induced by a given subset of classes. The SGFinder tool includes a graph representation of the class diagram and an efficient method of enumeration of induced subgraphs of a given size.

Ka-Yee et al. [6] use dynamic analysis and constraint programming to identify behavioral and creational patterns in source code. Using dynamic analysis, they reverse engineer the UML sequence diagrams from Java source code. Then they transform design pattern identification into a constraint propagation problem in terms of variables, constraints among them and their domains. They use a meta-model for the sequence diagram and transform behavioral and creational patterns into instances of the sequence diagram meta-model. Then, they compare two instances of sequence diagram meta-models: one for the pattern and the other for the design.

De Lucia et al. [9] propose an approach to identify behavioral design patterns. They combine static analysis with dynamic analysis. The static analysis extracts automatically from the design patterns the method calls. The dynamic analysis is performed through the automatic instrumentation of the method calls involved in the candidate pattern instances identified during static analysis. The dynamic information obtained from a program monitoring activity is matched against the definitions of the pattern behaviors expressed in terms of monitoring grammars.

In summary, none of the proposed approaches makes use of the semantic aspects in the pattern identification, in order to produce precise results for pattern identification. As explained in the introduction, such shortage does not distinguish between patterns that are structurally similar but semantically different.

3 The Semantics Based Design Pattern Detection Approach

Our design pattern approach tolerates structural and dynamic differences between the examined design and the identified pattern. In addition, it has the advantage of prioritizing the most important concepts representing the essence of a design pattern from its optional concepts. Furthermore, it can be used to identify the structure, the semantic and the behavior of the pattern. In fact, it is an adaptation of an XML document retrieval approach [13], tolerating structural variations of a pattern. Thus, we consider a design pattern as an XML query and the design as the target XML document where the pattern is retrieved.

The following sub-sections present first the XML retrieval technique, then the structural, the semantic and the behavioral identification.

3.1 XML Retrieval Technique

In our design pattern detection approach, we adopt the XML document retrieval technique proposed by Manning et al. [13]. In this work, the authors adapt the vector space formalism for XML retrieval by considering an XML document as an ordered, labeled tree. Each node of the tree represents an XML element. The tree is considered as a set of paths starting from the root to a leaf. In addition, each query is examined as an extended query that is, there can be an arbitrary number of intermediate nodes in the document for any pair of nodes in the query. Documents that are equivalent to the query structure by inserting less additional nodes are favored. The measure of the similarity of a path cq in a query Q and a path cd in a document D is the following context resemblance function [13]:

$$C_R(c_q,c_d)=\begin{cases}\dfrac{1+|c_q|}{1+|cd|} & \text{if } c_q \text{ matches } c_d\\ 0 & \text{if } c_q \text{ does not match } c_d\end{cases}$$

Where:

- $|c_q|$ and $|c_d|$ are the number of nodes in the query path and document path, respectively, and
- c_q matches c_d if and only if we can transform c_q into c_d by inserting additional nodes.

Note that the value of $CR(c_q, c_d)$ is 1 if the paths c_q and c_d in Q and D are identical. On the other hand, the more nodes separating the paths of Q and D, the less similar they are considered, i.e., the smaller their context resemblance value will be.

3.2 The Phases of Our Design Pattern Detection Approach

An overview of the principal activities performed by our approach is depicted in Figure 1 which shows: the general process of XML document retrieval and design pattern detection.

The first step consists in either re-engineering the code source or drawing the new design by ArgoUml. The second step transforms the XMI file corresponding to the design into an XML file valid according to the DTD of UML secondly; it decomposes the entire design obtained into sub-de while keeping all the structural information. This step aims to manage the time complexity of the pattern identification. Then, it identifies the structural features of the pattern through examining the class diagram (i.e., the classes, generalizations, aggregations, compositions, etc) [12] and checks if the pattern key methods are present in the design. It relies on methods name comparison. The next step analyses semantically the design to confirm the structural detection already performed in the previous step. This step allows design patterns identification without any confusion.

The final step identifies the dynamic aspect of the design pattern. It relies on the correspondence results determined by the previous steps. It also adapts an XML document retrieval approach to examine the conformity of the design behavior to that of the pattern. It supposes that the behavior is specified in terms of sequence diagrams. These different steps allow design patterns identification without any confusion.

In the following sub-sections, we will present the, briefly, structural and behavioral phases and then we will focus on the semantic phase. For more details on the behavioral phase, the reader is referred to [12].

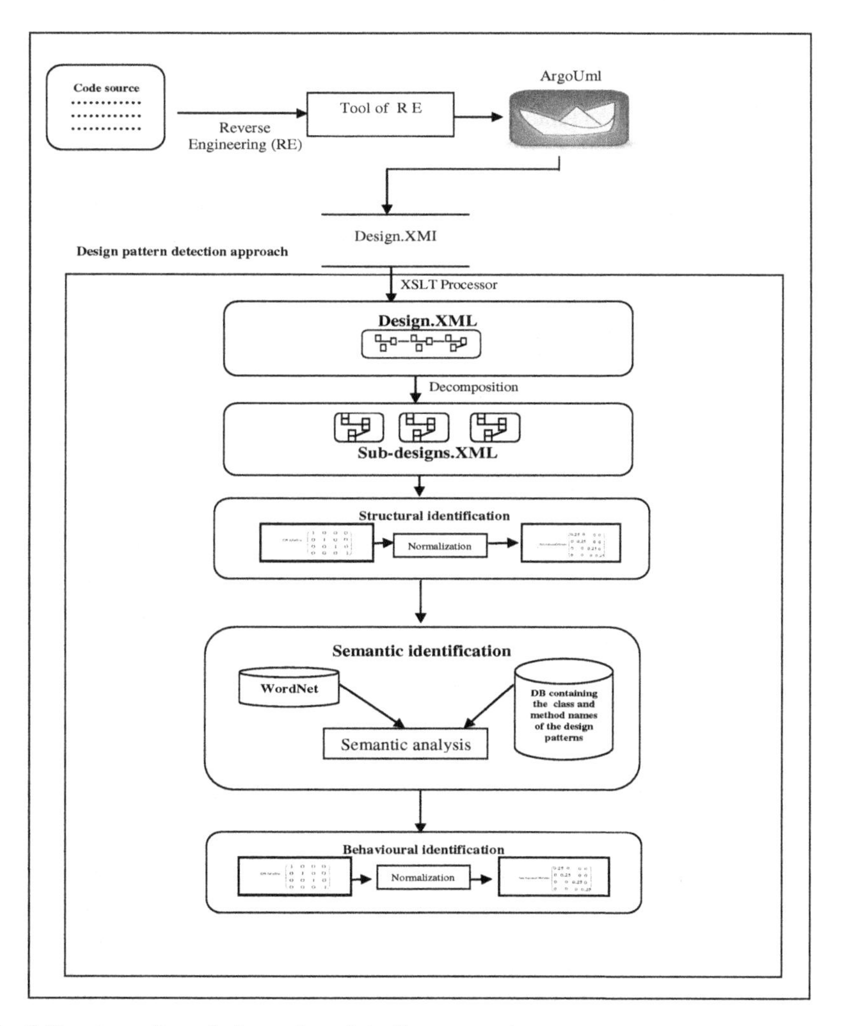

Fig. 1 The steps of our design pattern detection approach

3.2.1 The Structural Identification

For structural design pattern detection, we consider that a given pattern may be shown in many forms that differ from the core structure without losing the essence of the pattern. Thus, an exact pattern matching approach is inadequate. To handle the structural differences that may exist between the pattern and a design we adopt the XML retrieval technique. More accurately, we consider a design pattern as an XML query and the design as the target XML document where the pattern is searched.

The context resemblance function (CR) calculated in XML document retrieval, is based in general on an exact match between the names of the nodes in the query and the document paths. Nevertheless, for pattern detection, the nodes representing the classes are usually different in the pattern from those in the design. So, we first must calculate the resemblance values for the different matches between the class nodes in the pattern and those in the design. Then, we must consider the number of times a given match between two class nodes is used to calculate CR; and the importance of each relation in the pattern.

The structural resemblance between a pattern and a design starts by calculating the resemblance between each path of the pattern to all the paths in the design. The CRMatrix calculates the resemblance scores between the classes of the design and the classes of the pattern; this matrix sums up the values of the context resemblance scores for each class in the design with respect to a class in the pattern. This weighted sum accounts for the importance of the relations in the pattern; for instance, in the Composite pattern, the aggregation relation is more important than the inheritance relation. Afterwards, the scores of the CRMatrix are normalized with respect to the total number of classes in the design; the final matching results are collected in NormalizedCRMatrix whose columns are the classes in the pattern and whose rows are the classes of the design.

Now given the NormalizedCRMatrix, we can decide upon which correspondence better represents the pattern instantiation: For each pattern class, its corresponding design class is the one with the maximum resemblance score in NormalizedCRMatrix. Note that there might be more than one such class. This non-deterministic correspondence could be resolved through the semantic identification step.

3.2.2 The Behavioral Identification

The behavioral identification is based on the comparison of the sequence diagrams. In fact, to determine the behavioral resemblance between a design D and a pattern P, we extract the information from their corresponding sequence diagrams and we compare the ordered message exchanges for each pair of objects that were already identified as similar during the structural identification phase.

Thus, for each object O in a sequence diagram, its ordered message exchanges are represented through an XML path. Each node of these paths represents the type of the message (sent/received) along with the message being exchanged; this information allows us to derive a path where the edges have the same meaning: temporal precedence.

4 The Semantic Identification

After, identifying structurally the design pattern, we must analyze the semantic aspect of the identified pattern instance. Thus, we must determine the correspondences of the class and methods names corresponding to the pattern and to the structurally identified pattern instance. The semantic correspondence determination is an adaptation of the class names comparison criteria, proposed in [15].

In the following sub-sections, we define six types of relations between class names, and three types of relations between method names. All these relations express linguistic information between class and method names. Note that, they are obtained by using the WordNet dictionary [16]. Remark, also, that all the names are considered after eliminating the stop words such as "and", "the", "a"...

4.1 The Semantic Criteria

4.1.1 Class Name Correspondence Criteria

The following six criteria express the linguistic relationships between the class names of the design and the pattern:

- *Is_a kind_of(C_1,C_2)*: implies that there is a semantic relationship between two classes indicating that C_1 a type or a variation of C_2. *Example: Is_a kind_of (student, intellectual).*
- *Is_one way_ to (C_1,C_2):* implies that there is a semantic relationship between two classes. C_1 is one of several manners to do C_2. Example: *is_one_way_to (help, support).*
- *Synonym (C_1,C_2)*: implies that the name C_1 is a synonym of the name C_2. Example: *synonym (student, pupil).*
- *Inter_Def (C_1,C_2):* implies that the definitions C_1 and C_2 given by WordNet dictionary [16] have common words. The common words list is obtained after eliminating the stop words such as 'a','and', 'but',' how', 'or', and 'what'.
- *Def_Contain (C_1,keyword)*: implies that the definition of the name C_1 contain a certain keyword. Example: *Def_Contain (Paper, "Observation").*
- *Name_Includ (C_1,C_2):* implies that the name C_1 includes the name C_2. C_1 is a string extension of the name of the class C_2. *Example: Name_Includ* ("XWindow", IconXWindow).

Note that, the semantic criterion Is_a kind_of, Is_one way_to, and Synonym exist, already, in the WordNet dictionary [16].

4.1.2 Method Name Correspondence Criteria

The method name comparison criterion explicit the relation between the operations names belonging to the design and the pattern.

- *Synonym_Meth* (M_P, M_{C1}) :implies that the method M_P belonging to the pattern is identical or synonym to a method M_{C1} of the class C_1 belonging to the design , i.e., they have the same name or synonym names. Example: ("Build", Construct).
- *Inter_Def_Meth* (M_P, M_{C1}) : implies that the definition of the method M_P of the pattern and the definition of the method M_{C1} of the class C_1 belonging to the design have common words. *Inter_Def_Meth* ("handle", mouseDown).
- *Meth_name_Includ*(M_P, M_{C1}): implies that the name of the method M_{C1} of the class C_1 is contains the name of the method M_P belonging to the pattern. Example: *Meth_name_Includ* ("BuildPart", Build).

4.2 The Semantic Identification Rules

In our semantic identification step, we proposed 23 different rules for the identification of the different GoF design patterns [1]. These rules use the semantic comparison criteria, already presented. However, due to space limitations, we present, next, only few of them.

In this section, we will present the semantic identification rules for strategy, state, builder and observer design patterns. However, our approach is also applicable to the rest of GoF design patterns [1].

4.2.1 The Semantic Identification of the Strategy Pattern

The rule R 1 allows the semantic identification of the strategy design pattern (illustrated in Figure 2). Suppose that, the class C_1 plays the role of ConcreteStrategy1, C_2 plays the role of ConcreteStrategy2, C_3 plays the role of Context, and C_4 plays the role of Strategy. These different roles are obtained thanks to the structural identification.

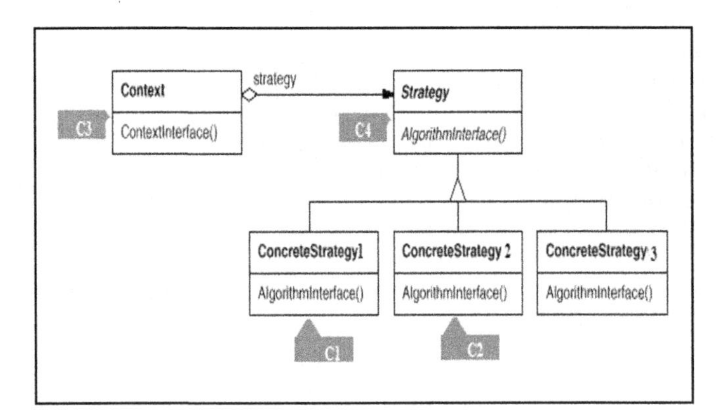

Fig. 2 Strategy design pattern

R1. If there is a set of classes, $C_1,\dots,C_n$, belonging to the design, such that C_1 plays the role of ConcreteStrategy1, C_2 plays the role of a ConcreteStrategy2, C_3 plays the role of Context, and C_4 plays the role of Strategy and *Is_one way_to* (C_1,C_3) or *Is_one way_to* (C_1, C_2) or *Def_Contain* (C_4, ("strategy","manner","style","mode", "way") or *Def_Contain*(C_4,synonym("strategy","manner","style","mode","way")) or *Name_Includ*("Strategy", C_4) or *Meth_name_includ*("Context",M_{C3})⇒the strategy pattern is detected.

Example:

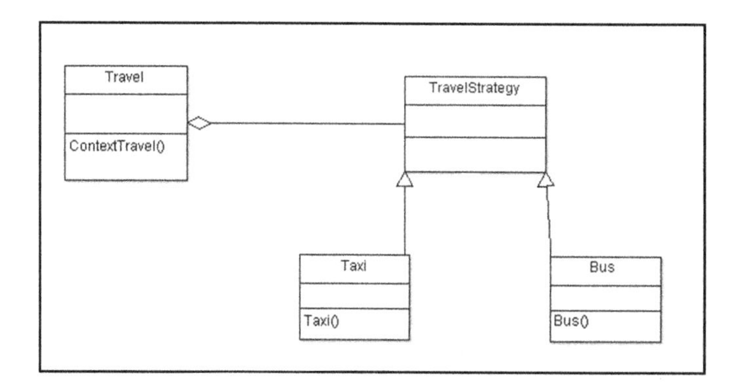

Fig. 3 A design example instantiating the strategy design pattern [17]

Let us illustrate the detection of the strategy design pattern [1] (Figure 2) and try to identify it in the design of Figure 3. After the structural identification, we found that the strategy design pattern is identified structurally and that the class *Travel* matches the *Context* class, the class *TravelStrategy* matches the *Strategy* class, the class *Taxi* matches the *ConcreteStrategy1*, and the class *Bus* matches the *ConcreteStrategy2*.

In order to confirm the identification of the design pattern, we are interested in the classes' semantic aspect. Thus, we apply the semantic rule of the strategy design pattern already presented. We found that: *Is_one way_to* (Taxi, Travel) and *Is_one way_to* (Bus,Travel), *Name_includ*("Strategy", TravelStrategy), *Meth_Name_Includ*("Context", ContextTravel).

4.2.2 The Semantic Identification of the State Pattern

The rule R 2 allows the semantic identification of the state design pattern (illustrated in Figure 4). As illustrated in this figure the class C_1 plays the role of *Context*, C_2 plays the role of *State*, C_3 plays the role of *ConcreteState1* and C_4 plays the role of *ConcreteState2*. These different roles are obtained thanks to the structural identification.

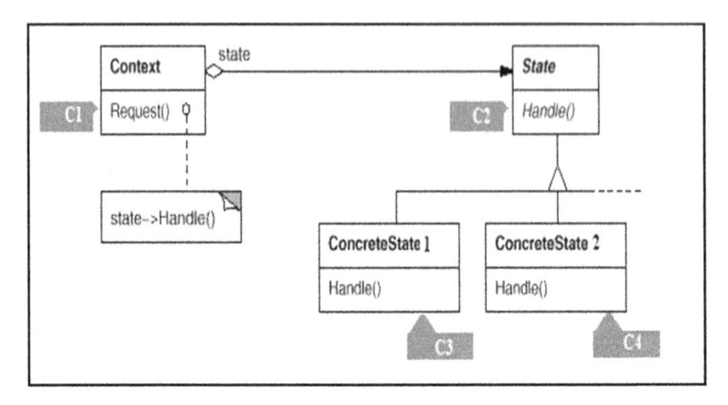

Fig. 4 The State design pattern

R2. If there is a set of classes, $C_1,\ldots,C_n$, belonging to the design, such that C_1 plays the role of *Context*, C_2 plays the role of *State*, C_3 plays the role of *ConcreteStat1* and C_4 plays the role of *ConcreteStat2* and *Is_a kind_of* (C_1,C_2) or *Is_a kind_of* (C_3, "change of State") or *Is_a kind_of* (C_4, "change of state") or *Name_Includ(*"State",C_2*)* or *Synonym_Meth*("Request",M_{C1}) or *Synonym_Meth*("Handle",M_{C2}) or *Synonym_Meth*("Handle",M_{C3}) or *Synonym_Meth* ("Handle",M_{C4}) or *Inter_Def_Meth*("Request",M_{C1}) or *Inter_Def_Meth*("Handle", M_{C2}) or Inter_Def_Meth ("Handle",M_{C3}) or Inter_Def_Meth ("Handle",M_{C4}) or *Meth_name_Includ* ("Request",M_{C1}) or *Meth_name_Includ* ("Handle",M_{C2}) or *Meth_name_Includ* ("Handle",M_{C3}) or *Meth_name_Includ* ("Handle",M_{C2}) $\Rightarrow$ the state pattern is detected.

As an example of the relation *Is_a kind_of* (C_4, "change of state"), we find Is_a kind_of (Start, "Change of State"), *Is_a kind_of* (Ending, "Change of State")

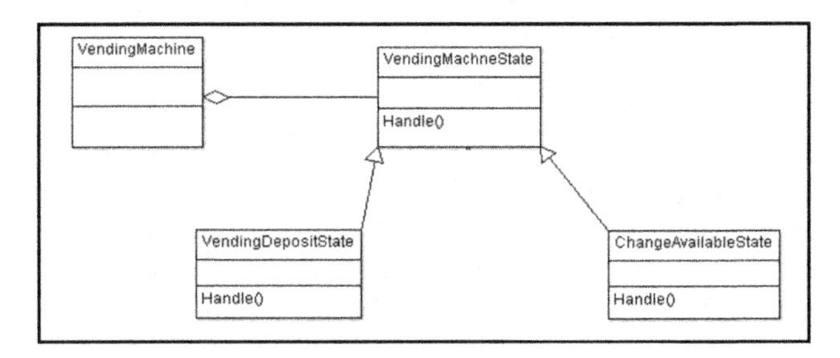

Fig. 5 A design example instantiating the state design pattern [17]

To illustrate the steps of our approach, let us illustrate the detection of the state design pattern [1] (Figure 4) and try to identify it in the design of Figure 5. After the structural identification we found that the state design pattern is identified and that the class *VendingMachine* matches the Context class, the class *VendingMachineState* matches the *State* class, the class *VendingDepositState* matches the *ConcreteState1*, and the class *ChangeAvailableState* matches the *ConcreteState2*.

Once a mapping between the classes of the design and those of the pattern is obtained by structural identification, we propose the verification of semantic aspect to confirm the identification of the suitable design pattern. Thus, we check the semantic rule of the state design pattern already presented. We found that: *Is_a kind_of* (VendingMachine,State), *Name_Includ*("State", VendingMachineState), *Synonym_Meth* ("Handle",Handle), *Synonym_Meth* ("Handle",Handle), *Synonym_Meth* ("Handle",Handle).

4.2.3 The Semantic Identification of the Builder Design Pattern

The rule R 3 allows the semantic identification of the builder design pattern (illustrated in Figure 6).

Suppose that, the class C_1 plays the role of Director, C_2 plays the role of Builder. These different roles are obtained thanks to the structural identification.

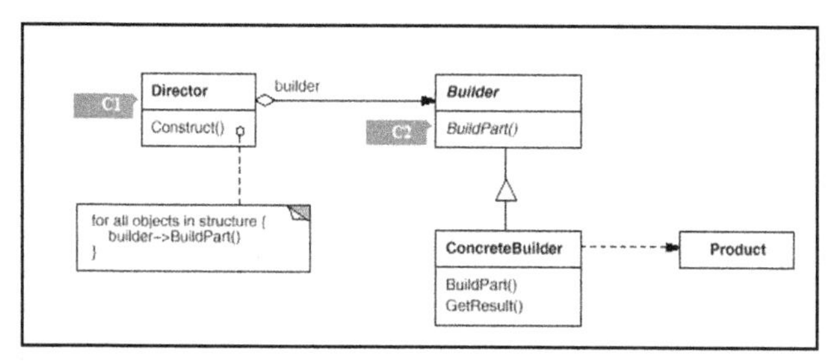

Fig. 6 The builder design pattern

R3. If there is a set of classes, $C_1,\ldots,C_n$, belonging to the design, such that C_1 plays the role of Director, C_2 plays the role of Builder and *Inter_Def*(C_1,C_2)={"Supervises","Construct", "Manufacture","Fabricate","Build"} or *Def_Contain*(C_1,"Director","Responsible", "Supervisor")) or *Def_Contain*(C_1, synonym("Director", "Responsible", "Supervisor")) or *Def_Contain*(C_2,"Builder", "Constructor", "Creator") or *Def_Contain*(C_2, synonym ("Builder", "Constructor","Creator")) or *Name_Includ*("Builder",C_2) or *Def_Contain*(C_1, synonym ("Constract", "Build")) and *Def_contain* (C_2,synonym("Constract","Build")) or *Synonym_Meth*("Construct",M_{C1}) or *Inter_Def_Meth*("Construct",M_{C1}) or *Meth_Includ*("Construct",M_{C1}).

Example

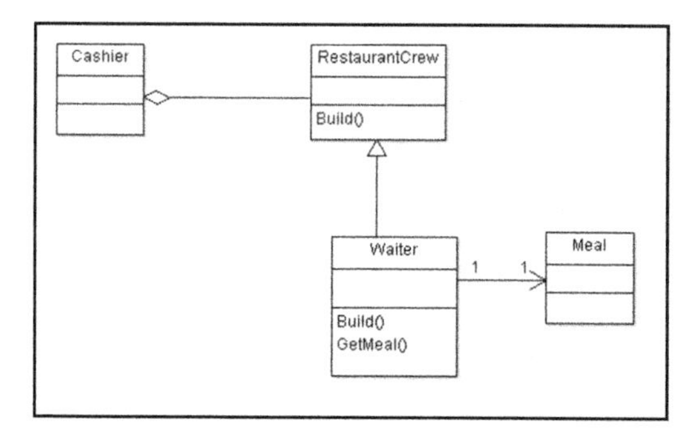

Fig. 7 A design example instantiating the builder design pattern [17]

To illustrate the steps of our approach, let us illustrate the detection of the Builder design pattern [1] (Figure 6) in the design of Figure 7. After the structural identification, we found that the Builder design pattern is identified and that the class *Cashier* matches the *Director* class, the class *RestaurantCrew* matches the Builder class, the class *Waiter* matches the *ConcreteBuilder*, and the class *Meal* matches the *Product* Class.

After this structural identification, we are interested in the classes' semantic aspect, in order to confirm the identification of the design pattern. Thus, we apply the semantic rule of the Builder design pattern. The result of this rule gives us that: *Def_Contain*(Cashier, "Responsible") *Def_Contain* (RestaurantCrew, "Buildment"), *Meth_Name_Includ* ("Build",BuildPart), *Meth_Name_Includ*("GetResault",GetMeal); *Meth_Name_Includ*("BuildPart",Build).

4.2.4 The Semantic Rule for the Identification of the Observer Design Pattern

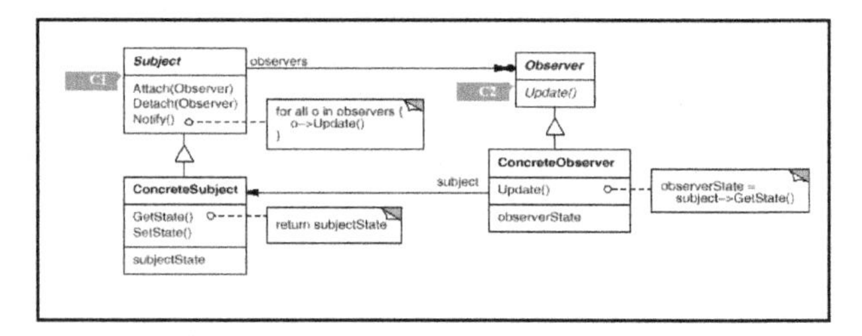

Fig. 8 The observer design pattern

The rule R 4 allows the semantic identification of the observer design pattern (illustrated in Figure 8).

R4. If there is a set of classes, $C_1,\ldots,C_n$, belonging to the design, such that C_1 plays the role of *Subject* and C_2 plays the role of *Observer* after structural identification.

Inter_def (C_1,C_2)={"Observe","Observation", "Observational"} or *Def_Contain*(C_1,"Subject") or *Def_Contain*(C_1,synonym("Subject")) or *Def_Contain*(C_2,"Observer") or *Name_Includ*("Observer",C_2) or *Def_Contain*(C_2, synonym("Observer")) or *Def_Contain*(C_1,"Observe")) or *Synonym*(C_2,"Observer") or *Synonym_Meth*("Notify",M_{C1}) or *Synonym_Meth*("Update", M_{C2}) or *Inter_Def_Meth* ("Notify", M_{C1}) or *Inter_Def_Meth*("Update", M_{C2}) or *Meth_name_Includ* ("Notify",M_{C1}) or *Meth_name_Includ* ("Update",M_{C2})

Example

Def_contain(Paper,"observations") and *Synonym* (reviewer, observer).

5 Example

To illustrate our approach, let us demonstrate the detection of design patterns in the design of Figure 9. After, the structural identification phase based on the XML retrieval technique, the following matrices are obtained. These matrices indicate the same values for the state pattern (Figure 2) and the strategy pattern (Figure 4). Note that these patterns have different intentions. In fact, the strategy defines a family of algorithms, encapsulate each one, and make them interchangeable. It lets the algorithm vary independently from clients that use it [1]. On the other hand, the state design pattern allows allow an object to alter its behavior when it's internal state changes. The object will appear to change its class [1].

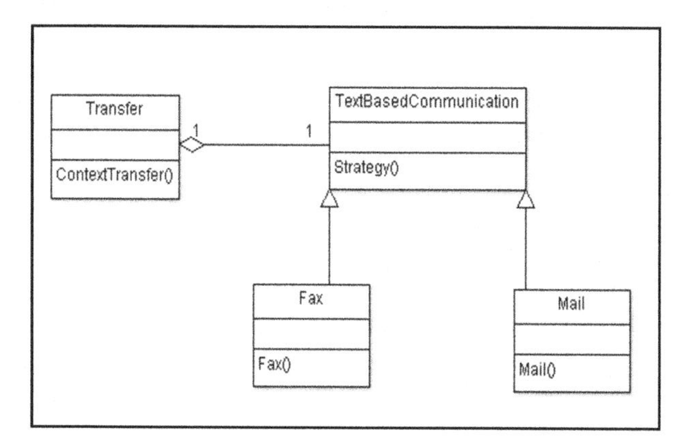

Fig. 9 A design example [17]

$$NormMatrix(design, State) = \begin{array}{c} \\ \text{TextBasedCommunication} \\ Transfer \\ Fax \\ Mail \end{array} \begin{array}{cccc} \text{State} & \text{CrtSt1} & \text{CrtSt2} & \text{Context} \\ 1.25 & 0 & 0 & 0 \\ 0 & 0 & 0 & 0.25 \\ 0 & 0.25 & 0.25 & 0 \\ 0 & 0.25 & 0.25 & 0 \end{array}$$

$$NormMatrix(design, Strategy) = \begin{array}{c} \\ \text{TextBasedCommunication} \\ Transfer \\ Fax \\ Mail \end{array} \begin{array}{cccc} \text{Strategy} & \text{CrtSg1} & \text{CrtSg2} & \text{Context} \\ 1.25 & 0 & 0 & 0 \\ 0 & 0 & 0 & 0.25 \\ 0 & 0.25 & 0.25 & 0 \\ 0 & 0.25 & 0.25 & 0 \end{array}$$

To summarize, we noted that with the structural identification, there are doubts that remain and we cannot decide if the strategy or the state pattern is the most suitable and which one must be detected in the design fragment. Thus, we deal with this problem, we perform the semantic analysis. We found that:

Is_one way_to (Mail, Transfer), *Is_one way_to* (Fax,Transfer), *Meth_name_includ*("ContextInterface", ContextTransfer) ⇒Thus, the appropriate design pattern is the Strategy.

6 Conclusion

The detection of patterns by considering the static and behavioral aspect is very important, but in case there are doubts, a third step that ensures the detection based on the semantics of class and method names is necessary.

We have presented in this paper an approach for pattern identification based on semantics. The approach considers a design pattern as an XML query to be found in an XML document representing a design. It uses an adapted context similarity function [13] to determine the correspondences between the classes of the design and those in the pattern. After structural detection we propose a semantic detection to confirm the identification of the design pattern using WordNet dictionary [16].

In our future works, we are examining how to use ontology in our semantic detection since WordNet dictionary is not complete.

References

1. Gamma, E., Helm, R., Johnson, R., Vlissides, J.: Design patterns: Elements of reusable Object Oriented Software. Addisson-Wesley, Reading (1995)
2. Tsantalis, N., Chatzigeorgiou, A., Stephanides, G., Halkidis, S.T.: Design pattern detection using similarity scoring. IEEE Transactions on Software Engineering 32(11) (2006)
3. Gueheneuc, Y., Antoniol, G.: DeMIMA: A Multilayered Approach for Design Pattern Identification. IEEE Transactions on Software Engineering (2008)
4. Dong, J., Sun, Y., Zhao, Y.: Design pattern detection by template matching. In: SAC 2008 (2008)
5. Belderrar, A., Kpodjedo, S., Guéhéneuc, Y., Antoniol, G., Galinier, P.: Sub-graph Mining: Identifying Micro-architectures in Evolving Object-oriented Software. In: 15th European Conference on Software Maintenance and Reengineering, CSMR 2011, pp. 171–180 (2011)
6. Ka-Yee Ng, J., Gueheneuc, Y.: Identification of behavioural and creational design patterns through dynamic analysis. In: The 3rd International Workshop on Program Comprehension through Dynamic Analysis, PCODA, pp. 34–42 (2007)
7. Lee, H., Youn, H.: A design pattern detection technique that aids reverse engineering. The International Journal of Security and Applications (2008)
8. Albin Amiot, H., Cointe, P., Guéhéneuc, Y.G.: Un meta-modele pour coupler application et detection des design patterns. L'objet N°8, pp. 1–18 (2002)
9. De Lucia, A., Deufemia, V., Gravino, C., Risi, M.: Behavioral Pattern Identification through Visual Language Parsing and Code Instrumentation. In: European Conference on Software Maintenance and Reengineering, CSMR 2009, pp. 99–108 (2009)
10. De Lucia, A., Deufemia, V., Gravino, C., Risi, M.: Improving Behavioral Design Pattern Detection through Model Checking. In: 14th European Conference on Software Maintenance and Reengineering, CSMR 2010, pp. 176–185 (2010)
11. Arcelli, F., Tosi, C., Zanoni, M., Maggioni, S.: JADEPT: Dynamic analysis for behavioral design pattern detection. In: 4th International Conference on Evaluation of Novel Approaches to Software Engineering, ENASE 2009, pp. 95–106 (2009)

12. Bouassida, N., Ben-Abdallah, H.: Structural and Behavioral Detection of Design Patterns. In: Ślęzak, D., Kim, T.-h., Kiumi, A., Jiang, T., Verner, J., Abrahão, S. (eds.) ASEA 2009. CCIS, vol. 59, pp. 16–24. Springer, Heidelberg (2009)
13. Manning, C., Raghavan, P., Schütze, H.: An introduction to information retrieval. Cambridge University (2008)
14. Blondel, V.D., Gajardo, A., Heymans, M., Senellart, P., Van Dooren, P.: A Measure of Similarity between Graph Vertices. In: Applications to Synonym Extraction and Web Searching. SIAM (2004)
15. Bouassida, N., Ben-Abdallah, H.: A New Approach for Pattern Problem Detection. In: Pernici, B. (ed.) CAiSE 2010. LNCS, vol. 6051, pp. 150–164. Springer, Heidelberg (2010)
16. Fellbaum, C.: pp. 665-670. Elsevier (2005) (Online), http://wordnet.princeton.edu/
17. Duell, M., Goodsen, J., Rising, L.: Examples to Accompany: Design Patterns Elements of Reusable Object-Oriented Software (1999)

Enterprise Architecture: A Framework Based on Human Behavior Using the Theory of Structuration

Dominic M. Mezzanotte, Sr. and Josh Dehlinger

Abstract. Organizational knowledge provides the requirements necessary for effective Enterprise Architecture (EA) design. The usefulness of EA processes depend on the quality of both functional and non-functional requirements elicited during the EA design process. Existing EA frameworks consider EA design solely from a techno-centric perspective focusing on the interaction of business goals, strategies, and technology. However, many organizations fail to achieve the business goals established for the EA because of miscommunication of stakeholder requirements. Though modeling functional and non-functional design requirements from a technical perspective better ensures delivery of EA, a more complete approach would fully take into account human behavior as a vital factor in EA design. The contribution of this paper is an EA design guideline based on human behavior and socio-communicative aspects of both stakeholders and the organization using socio-oriented approaches to EA design and modeling schemes.

1 Introduction

Technological developments and acceptance of that technology by stakeholders drive social, economic, and political transformation within an organization [3][4][27]. In Enterprise Architecture (EA), gains in organizational efficiency, effectiveness, employee productivity, and alignment of the organization's strategic business plan with it's information technology (IT) capabilities are lost if users do not accept the rationale for or fail to perceive the benefits in the usefulness, ease of use, and need for technology [4][13]. EA failure occurs frequently when project deliverables are late, over-budget, and fail to meet either stakeholder expectations or those of the organization paying for the effort. The cost of failed IT projects world-wide range annually into the billions of dollars for both the private and public sectors [9][11][15][17][30].

Dominic M. Mezzanotte, Sr. · Josh Dehlinger
Department of Computer and Information Science
Towson University
e-mail: {dmezzanotte,jdehlinger}@towson.edu

R. Lee (Ed.): Software Engineering Research, Management and Appl. 2012, SCI 430, pp. 65–79.
springerlink.com

The causes for failure are difficult to explain ranging from complexities found in the interactions of technology to the tools (frameworks, models, software, procedures and processes) used to transform collected data into information [8][18][27]. Part of the complexity lies in the underlying dynamics associated with stakeholder behavior and their interaction with and reaction to the EA and technology.

Understanding human behavior is based largely on how a person perceives and thinks about a given circumstance. The difficulty here is that stakeholder behavior in EA does not always conform to management desires, expectations, or directives and are pessimistic, perceptive, and capable of acting on their own satisfying their own goals and objectives when confronted by organizational transformation brought about by EA.

EA introduces new processes and procedures into the workplace and as a result causes organizational transformation and the surrounding social structure with new roles, responsibilities, and duties assigned to stakeholders [4][5][27]. We can then posit organizational change as emerging from an unpredictable interaction between technology and new emotions and feelings (i.e., behavior changes) exhibited by stakeholders. The behavioral changes can then be tied to the impact of new technology caused by a perceived diminution of power and influence within the organization manifesting itself in behavioral patterns that are not aligned with achieving organizational EA goals and objectives [14][22][23]. Simply stated, stakeholders may be more interested in their own parochial set of goals and objectives than in those of the organization. In general, these behavior patterns can constrain and limit human action, stifle stakeholder innovation and creativity, and more importantly affect the quality of requirements provided by stakeholders. Thus, organizational transformation and resulting behavior changes require EA management to exercise psychological and sociological processes that go beyond technical and business knowledge that monitor and govern stakeholder action [5].

The interactions of human resources involved in EA, the stakeholders and their behavior, play a critical role in the success or failure of EA [27]. The manner in which EA planning and development takes place can therefore affect stakeholder willingness to contribute to and accept new technology. For example, if new technology is introduced unexpectedly into the organization without any input from stakeholders, it may be accepted, rejected, or modified by stakeholders to satisfy their own personal goals and objectives. From an analytical perspective, these personal goals and objectives may:

- Be contrary to those of the organization
- Represent the self-interests of the stakeholder
- Limit stakeholder innovation and creativity
- Include the perceptions and be influenced by the behavior of other group members
- Affect their capacity to act within and for the organization or to undercut its policies and procedures
- Pose an influence on and perhaps a major threat to EA success [12][23][27].

Most EA failures are because of internal barriers to EA, most of which are human and assigned, mistakenly, to "poor architecture" [4][8]. Identifying the root-cause for "poor architecture," we find the attribution to be the by-product of misunderstood and/or miscommunicated EA design requirements [5][27]. In this context, requirements represent knowledge creation elicited from the accumulation of embedded institutionalized organizational knowledge learned and retained by stakeholders over some period of time. This knowledge falls into two major categories:

- Explicit knowledge (that which is usually documented, well known within the organization and is, in most cases, easily verifiable and reliable)
- Tacit knowledge (that which is more elusive to obtain, typically is undocumented, and known only to an individual or select group(s) of people).

Together, explicit and tacit knowledge defines the necessary components needed to determine the functional and non-functional requirements needed for EA [6][26][33]. The critical nature and role of this knowledge in EA design changes the very complexion and significance in evaluating the elicitation process and thus poses a dilemma in determining how stakeholders create, process, produce, and, more importantly, provide organizational knowledge as input to EA design.

Eliciting stakeholder requirements is in reality a dynamic, recursive process that re-conceptualizes organizational knowledge in that the quality of that knowledge can be directly linked to stakeholder and organizational behavior and the organizational environment [5][27]. If stakeholders are committed to EA, the elicitation process will yield legitimate design requirements. If they are not committed to the EA, the EA has a good chance to fail. This link thus ties together the importance of human behavior and social structure recognizing both as significant drivers in the acquisition of design requirements needed for EA.

The motivation for this research progresses our earlier work [24][25] on human and organizational behavior and their intersection with the complex multi-faceted system/entities of EA design and organizational transformation. We propose a framework that: takes into account human and organizational behavior as key facets of EA design, identifies the risks to management resulting from behavioral factors, and postulates a process that mitigates and/or reduces significantly the surrounding uncertainties inherent in the design and implementation of purely techno-centric and techno-oriented design and modeling processes. This paper asserts that a better EA solution includes stakeholder and organization behavioral as major components of the EA development process focusing on:

- The impact of human behavior as an input to EA design
- Exploration of the recursive nature of human behavior in formulating and designing EA
- Development and use of an information gathering process to better ensure quality of EA design requirements.

From this position, we can explore and observe the interactions, intersection, and behavior of human resources as vital in the elicitation and validation of design requirements and thus better ensuring EA viability. Our approach utilizes aspects of

Giddens' *Theory of Structuration* and its application in the realm of technology as it relates to and can be used as a lens guiding EA design and implementation.

The remainder of this paper is as follows. Section 2 examines current EAFs and modeling schemes for their approach to stakeholder behavior as an input to, and a reaction from, the development of an EA [26][31]. Section 3 describes the inclusion of a behavioral and organizational theory, the *Theory of Structuration* [12], as a lens by which the development of an EA can be used to understand the importance of stakeholder behavior in EA design and mitigate EA failure. Section 4 provides a brief and preliminary description of an architectural framework that takes into account human and organizational behavior that can be used to either enhance existing EA design and modeling methodologies. Section 5 provides a brief discussion, some concluding remarks, and our plans for future research directions in enhancing EA design.

2 An Analysis of Existing Enterprise Architecture Frameworks

The first documented EA framework (EAF), the Zachman Framework for Enterprise Architecture (Z|AF), is a taxonomy-oriented framework for development of large, complex systems [32] and uses a matrix of cells and rows to classify design requirements based on stakeholder roles and expectations [36]. This framework was followed shortly thereafter by other ontological-oriented EAFs such as The Open Group's Architecture Framework (TOGAF) [28], the Federal government's Federal Enterprise Architecture Framework (FEAF) [2] and the Department of Defense Architecture Framework (DoDAF) [1][32].

Existing EAFs are techno-centric with each providing a comprehensive mechanism for handling large volumes of complex and inter-dependent system and subsystem requirements producing as a process output an EA plan (EAP) [26][28]. Current state-of-the-practice EAPs formulate an EA aimed at maintaining business continuity and aligning an organization's strategic business plans (goals, vision, strategies, and governance principles), business operations (business vocabulary, organizational structure, people, procedures, processes, and information), with it's IT infrastructure (e.g. hardware, software, processes, procedures) and resources [9][13][26]. The approach follows a macro-oriented, high-level abstraction of EA design artifacts and requirements.

Conversely, the inherent weakness of each EAF centers on the techno-centric solutions they formulate towards producing a desired set of technical deliverables for the EA [9][26]. This satisfies the macro-oriented high-level abstraction needed for EA with the strategic EAP identifying in detail the proposed organizational structure, business processes, desired information systems (IS), design requirements, implementation plan, and associated IT infrastructure. However, the processes discount the importance of the intersection of technology with human behavior and their respective effect on the quality of the work effort, the EA design requirements.

For example, the key element around which all design activity takes place in TOGAF Version 9.1, the Application Development Method (ADM) [29], describes a comprehensive step-by-step approach to develop EA [29]. It provides for the identification and management of key stakeholders from a purely technical perspective. Stakeholder roles, responsibilities, and contribution to EA are identified through TOGAF's ADM based on what is termed "Stakeholder Management." The process consists of four concepts: Stakeholders, Concerns, Views, and Viewpoints. The process essentially identifies who will be involved and needed in EA design [29]. The process asks several questions regarding stakeholder role, decision-making, and resource control, etc. Though these questions by themselves sound relevant, the process itself fails to ask questions that would improve good decision-making and problem-solving such as "why" and "why not?"

Taking a page from Kepner-Tregoe [16], we can examine the selection of project stakeholder from two points of view. For example, asking the question "why" or "why not" to the stakeholder selected for a key role in the EA might provide better insight into the value and contribution that stakeholder might make to the EA which might then disqualify that person. In both scenarios, asking these kinds of questions allows EA management and the Enterprise Information Architect (EIA) [24] to place a value on the provided answers and assess the potential contribution or lack thereof for every aspect of EA design.

From EA's beginnings, little effort has been expended to provide for enterprise engineering, enterprise modeling or inclusion of organizational and human behavior to support EAFs. The iterative process used in EA design, and corresponding EAFs, recognizes the critical nature of and depends on the elicitation and validation of stakeholder input (requirements). Coupled with the recursive nature and cognitive aspects of stakeholder behavior, the complexity of the design process becomes more difficult to solve as human behavior now becomes another independent variable in this process. Failure to correctly elicit, document, and communicate these requirements back to the stakeholders to verify correctness, and obtain concurrence most certainly leads to failed EA.

There have been many proposed solutions to assess the validity of EAF design requirements [10][15][28][35]. The solutions also include a plethora of modeling languages such as AORML and TROPOS which address systems design (SD) and systems requirements (SR) and modeling schemes such as *i** [35] which discuss the need for and application of ***soft people*** skills, none adequately address the sociological factors that impede EA design requirements.

The regimen imposed by existing EAFs limits stakeholder input due to organizational rules, policies and assumptions creating an environment that is reluctant to contribute effectively to problem-solving [21][27]. The EAFs provide little provision for incorporating patterns of stakeholder behavior and organization (i.e. communication) which we believe is a fundamental ingredient to manage human behavior and monitor and govern EA.

Communication mechanisms allow management to exercise EA governance and to inform stakeholders of subsequent organizational change caused by the EA [21].Therefore, the EA should be predicated upon a communications path, both vertical and horizontal, that recognizes the cognitive effect of human behavior, the impact of technology, and the actions and relationships of those entities on and within the social structure of the organization (see Figure 1). The communications path should foster a participative EA design that encourages ownership by EA stakeholders. In effect, a *socio-communicative* path enhanced with a feedback mechanism that provides a means of maintaining a state of homeostasis offering a channel for stakeholders to exchange ideas, know-how and knowledge benefitting the organization.

Perhaps a more important realization recognizes that human behavior is a combination of distinctly different cognitive processes and mental phenomena, and that strategic problem-solving is better served by an EAF and modeling technique that incorporate human behavior and an effective communications path in it's approach to EA. This approach permits a more holistic, dynamic, and broader approach that would produce better EA design by providing the ability to explore and integrate different human and organizational models that simulate realistic approaches to problem-solving.

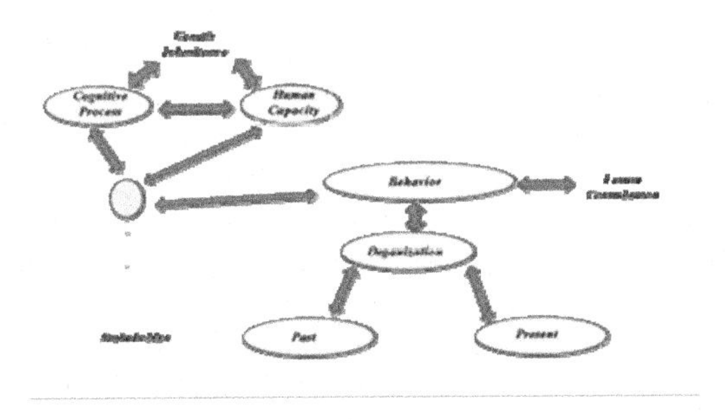

Fig. 1 The Cognitive Nature of Stakeholder Behavior.

3 Enterprise Architecture and the Theory of Structuration

EA emerges from many definable and fairly easily created business circumstances. However, it does take a well designed plan for developing the EA and persistent effort over an extended period of time that is frequently punctuated by instantiated evolution to accomplish.

EA design and modeling schemes pervade the IT world all aimed at improving operational efficiency, effectiveness, and producing the real reason for EA, profit at low cost [18][26]. However, the organizational goals and objectives desired of EA are not always those of stakeholders responsible for doing work and as such pose a paradox and a dilemma associated with this subject.

Existing EAFs and modeling schemes follow generally accepted software engineering (SE) and requirements engineering (RE) practices with the expected outcome an EAP providing the building blocks for EA [26][31][34]. Today, stakeholders frequently question, either covertly or overtly, new technology in relation to how it might affect the environment in which they function and to how it will affect their status within the organization and their new roles, responsibilities, and duties [27]. In this context, stakeholders may accept, reject, or modify the technology to fit what they perceive to be their desired roles [27]. In this same context, stakeholders may either abuse technology or use it as intended. This raises questions about how EIAs must elicit, verify, validate (model) and evaluate the quality of each requirement and subsequent transformation into a technical design specification. From this perspective, we believe EIAs must take into account stakeholder behavior and how the unanticipated use of proposed technology might affect EA design throughout the EA design effort [9][27]. Therefore, the EIA must include design solutions that identify and handle misuse of the technology.

Organizations can be viewed as a complex network of technical relations and systems/entities for getting institutionalized work done (output) [4]. We can represent the organization as a system/entity consisting of a series of independent variables such as management attitudes and behavior (including that directed at employees), organizational standards consisting of policies and rules, structure and operational environment, etc., [12][14][22][23]. Technology can be viewed as a constant variable consisting of the arrangement of people producing work, application software, etc., [3].

Organizational behavior is based on a set of standardized rules, procedures, processes, and systems (collectively referred to as rules) [12][22][23]. These rules constitute a set of coordinated and controlled activities with institutional work (i.e., output) produced from complex networks of technical relations that span organizational boundaries [27]. The rules and activities then become an integral part of prevailing social behavior within the organization and are subsequently built into the society as reciprocated interpretations [5][10]. This rule-based environment inhibits and limits a dynamic view of the organization and discourages participation and, in general, is not conducive for EA stakeholders to be innovative, creative, and therefore more receptive to change [12][27]. We consider the recursive nature of people and organizations as postulated by Giddens' *Theory of Structuration* [12] as a theoretical lens to align EA and IT with a socio-communicative framework that takes into account both stakeholder and organizational behavior.

Until recently, the *Theory of Structuration*, has been a theory based on the social sciences, human action, and organizational structure paying little attention to IT. However, the application of the *Theory of Structuration* to IT lends itself as a design tool for EIAs to better understand stakeholder behavior and the conceptual impact of technology on organizational behavior. Orlikowsi, one noted advocate for using the *Theory of Structuration* in IT development and deployment, proposed the *Structurational Model of Technology* (SMT), as a means to understand how technology affects organizations and vice versa [27]. *SMT's ap*proach centers on two concepts - the *Duality of Technology* and the *Interpretive Flexibility* of technology [27]. The former posits that the socially created view and the objective view of technology is not exclusive but intertwined and are differentiated because of the temporal distance between the creation of technology and usage of the same. *Interpretive Flexibility* on the other hand defines the degree to which users of a technology are engaged in how it is built and used (physically and/or socially) [27].

In summary, the theoretical premise of the *Theory of Structuration* [12] and *SMT* [27] recognizes and highlights those factors that influence organizational structures, technology and human action as intertwined activities/entities such that each is continually reinforced and transformed by the other (a recursive process). A logical conclusion can therefore be made that the veracity of design requirements demands a dynamic theory of social and institutional order to ensure quality EA design.

4 A Paradigm for Modeling Enterprise Architecture

In earlier work, [24][25], we identified and addressed several of the causal factors leading to EA failure. From that work, we propose a solution and approach where management, the EIA, and key stakeholders collectively define, establish, and execute a management system and reliability plan that addresses quality from a human behavioral point-of-view recognizing and handling both negative and positive stakeholder behavior. The basic framework consists of three distinct processes:

- Design and use of an Architectural Design Plan (ADP) that describes the conduct of the EA based on an analysis of the existing organizational environment and management style
- Establishment and use of a Socio-Communicative process that handles all aspects of inter and intra organization communication
- Implementation of a Requirement Process Chain for elicitation and validation of EA design requirements.

The on-going development of the framework provides a holistic procedure based on psychological and sociological approaches aimed at both motivating and acknowledging positive human behavior and incorporating checkpoints to anticipate and provide a paradigm to resolve and mitigate negative human behavior and influences on EA.

Historically, social theorists assert that top-management behavior permeates through all layers of the organization influencing the organization's work environment and social structure [12][14][22][23]. From this perspective, we can state that organizations, represented by its management behavior, attitudes, rules, and policies maintain a deeply ingrained mechanistic view of technology stemming from a utilitarian economic emphasis on operational efficiency. Stakeholders, on the other hand, are purposeful systems which exhibit will which may act in concert with or oppose organizational goals and objectives and that the addition of technology to this equation [4][12][20][22]:

- Compels stakeholder action to accept, reject, or modify potential EA solutions
- Influences their behavior and attitudes towards the organization
- Affects the feedback loops and selection mechanisms for managing, monitoring, and governing EA design activities.

These then form the basis for a two phased analytical approach to problem-solving wherein:

- A communications forum establishes an effective path for information sharing and the transfer of explicit and tacit knowledge throughout the organization

- An agreed-upon process for eliciting EA design requirements that better ensures collection and validation of design requirements.

Coupling these two processes provide insight and a more meaningful and cogent solution to many EA failure factors.

One of the often cited contributors to EA failure is the lack of or poor leadership, often attributed to both organizational management and the role played by and activities of the EIA. However, leadership, though vital to overall EA direction, may not be the major cause of EA failure [19]. In most organizations, human behavior follows norms, rules, and practices established by the organization either by dictate or acquired and institutionalized over time. However, another facet to this behavior is that actually demonstrated by the employee which may or may not conform to organizational desires, goals, and objectives [14][27]. Our goal, therefore, is to establish an environment where stakeholders are willing to share knowledge needed and essential to EA success. In effect, an environment that promotes, encourages, and fosters user acceptance and ownership of the EA. Therefore, an EA environment must be established that transcends technical and business issues and includes psychological and sociological concerns focused more on the recursive nature and human aspects of the EA.

To illustrate the importance of human behavior, the behavioral questions from this perspective organizations are faced with are:

- How will the introduction of new technology affect the organizational structure, environment and transformation
- What possible changes in human behavior may result from new technology
- Will the change meet with stakeholder resistance and/or a reluctance to adhere to new policies and procedures?

Each of these concerns represents human behavior in one manner or another and therefore must be taken into account to assess the impact and influence EA brings to the enterprise. People view technology from different perspectives, reacting to it accordingly, and for several legitimate reasons:

- Fear of job loss and job security
- A perception of loss of status, role, duties, and responsibility within the organization
- The need to learn new procedures and processes
- Belief and feeling that the employer no longer cares about the employee.

These factors attribute to EA failure as detailed in our earlier work [24][25]. The primary issue then confronting the EIA and organization is how to avoid these factors and prevent EA failure.

4.1 The Enterprise Architectural Design Plan

In a typical EA, it is not a matter of choosing which requirements to meet but of trying to meet all practical requirements. This first step proposed for EA is an

Architectural Design Plan (ADP) consisting of two components: a Development Plan (DP) and a Control Plan (CP) (see Figure 2).

The ADP documents and establishes how the overall conduct of the EA is to be progressed, stakeholders selected and assigned to the project, the kinds of procedures to be used in eliciting design requirements, the communications and feedback loop(s) needed to verify design requirements, and the measurement, monitoring, and governance techniques needed to ensure the validity of the design requirements. The primary purpose of the plan is twofold:

- Provide the mechanism for the EIA to learn the existing organizational environment and identify areas of potential concern
- Provide the basic scheme for eliciting information (requirements) from which to design the EA.

This process details an excellent opportunity for the EIA to learn not only what needs to be done but also who is to participate along with their personalities, how the project is to be managed and governed, and why it needs to be done.

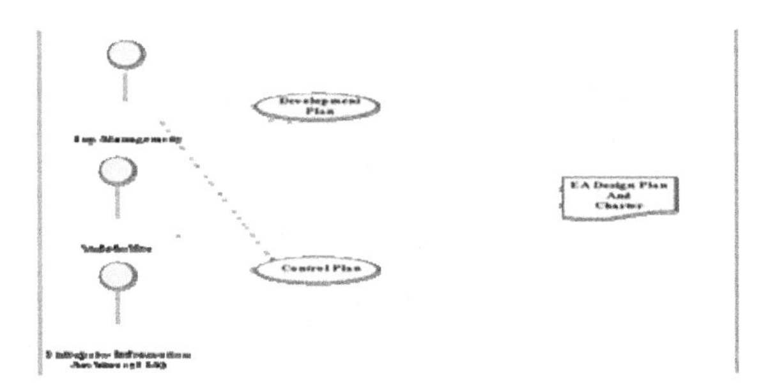

Fig. 2 The Enterprise Architectural Design Plan.

The second step in this process defines and describes the specific design handles used to ensure stakeholder requirements are met. Organizational capabilities, organizational reaction to nonconformance and then exact critical in-design parameters that control the quality attributes of the design are documented and assessed to ensure that exact stakeholder expectations are met.

This step is performed simultaneously with the ADP and formulates and establishes a communications scheme that will be used throughout the project life cycle.

4.2 A Socio-communicative Approach to EA Design

Technology alone causes significant changes to organizational philosophy, social structure, and environment coalesced into three crucial artifacts affecting stakeholder behavior:

- The magnitude of noxiousness of a depicted event
- The probability of that event's occurrence
- The efficacy of a response.

Investigation into the relevance of these components leads to potential solutions easily addressed using communication as a process to initiate a corresponding cognitive appraisal of stakeholder behavior and technology.

The communication process focuses on mechanisms for motivating stakeholders and that, at the same time, manipulates and mediates attitudes towards change. In this context, communication is a choice and therefore requires encouragement and willingness on the part of stakeholders to participate to be successful and thus become a stakeholder motivator.

The second process begins by putting in place a communications mechanism and path extending from both a top-down and a bottoms-up scheme that fosters an environment that encourages:

- Sharing of ideas (□brainstorming□) that facilitate decision-making, creative and critical thinking, and problem-solving
- Explaining decisions and providing an opportunity for clarification
- Sharing responsibility for decision-making and implementation
- Providing specific instructions and closely supervising and rewarding performance (governance)

The process encompasses three aspects of communications (see Figure 3): Disclosure, Analysis, and Feedback (DAF).

Fig. 3 The Disclosure, Analysis, and Feedback (DAF) Communications Process Loop.

DAF establishes, as a first step, an on-going, open-ended forum for transmission and reception of all ideas, comments, observations, and questions ("brainstorming") related to the EA while identifying potential surrounding internal and external influences that might affect the EA. The procedures for EA management, measurement, and governance are instantiated, documented and distributed to all stakeholders, and initiated before EA design begins.

The process defines how the EA will be conducted, documented, continued, and maintained throughout the EA life-cycle. The second step requires top-level management to publish a description of what EA is, describing its purpose, the reasons for, and the role EA has within the organization detailing an initial list of EA goals and objectives including the magnitude, scope, and boundaries for the EA. Depending on the complexity of the anticipated work, plans for scheduling any education and training of stakeholders should be included and distributed to all project stakeholders.

Education, training, and the communications channel can also be viewed as stakeholder motivators. The third step defines and establishes the communications process describing the:

- Formats for all EA design and implementation correspondence such as distribution lists, review meetings, and progress reports
- Procedures to be used in the collection and feedback of EA requirements
- Rules, policies, and practices to be used to manage, monitor, and measure EA design and progress
- Mechanisms for administering and managing the EA governance
- Procedures providing for input to the EA design suggesting innovative and creative ideas, opinions, and endogenous and exogenous factors that might influence the decision-making process.

The goal of this proposed framework is to create an environment and as architecture that provides best chance for success as well as the most adaptable, practical solution for the future and which aligns strategic business with IT plans.

4.3 Requirement Process Chain

The third process of our paradigm, the Requirement Process Chain (RPC), consists of four distinct phases:

- The Requirement Collection Process
- Management, Monitoring, Measurement, and Governance
- Quality Assurance and Reliability
- Failure Risk Analysis and Effects Mode.

The Requirement Collection Process details the methods to be used such as interviews, questionnaires, surveys, and analysis of existing documentation all aimed at the elicitation of explicit and tacit stakeholder knowledge (requirements) and describing how the information will be verified, validated, and accredited (see Figure 4).

The second step of this process describes how the overall EA will be managed, monitored, measured, and governed such that all stakeholders are aware of the procedures and feedback loops are to be employed. This phase assigns specifics related to stakeholder requirements are to be handled and modeled to ensure the process is capable of meeting stakeholder needs and requirements.

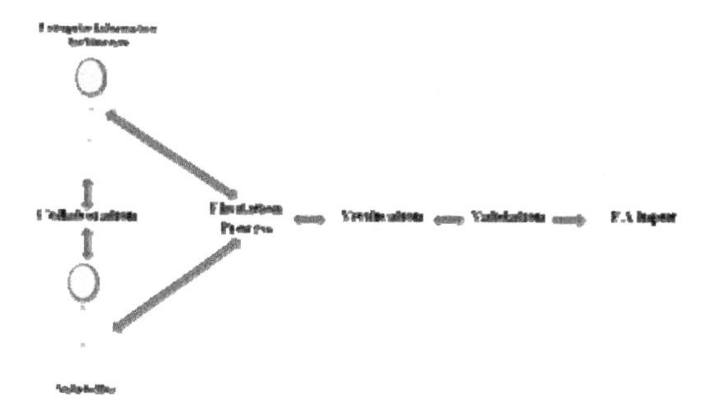

Fig. 4 Requirement Collection Process (RCP).

The Quality Assurance and Reliability phase ensures proper communication takes place between all parties affected by the new design. Results of tests, models, human-computer interface analysis in the traditional technical behavioral sense, documentation, and stakeholder acceptance and sign-off procedures will be distributed to top-management through the DAF communications channels.

The Failure Assessment and Effects Phase, a "brainstorming" event, examines and analyzes every possible failure mode for the process under examination in order to understand the consequences of process failure from the stakeholder perspective. The goal of this phase is to provide a risk-analysis for the failed process and a plan to correct the deficiency in system or software design and development.

5 Discussion, Future Direction and Closing Remarks

EA and the introduction of new and/or enhanced technology into an organization often results in a sociological and a political change in the hierarchical structure of the enterprise. This is evidenced by a dynamic shift in internal and perhaps external perceptions of the organization. Stakeholder roles, responsibilities, and duties invariably change because of new rules, policies, procedures and processes introduced by technology into the organization. Therefore, the manner in which the EA design takes place can seriously affect acceptance and/or rejection of the EA by different stakeholders.

We plan to advance our work into gaining a better understanding of human behavior and the influence it has on EA either positively or negatively to provide a better platform to manage and govern the design process. The reliability and management system based on human behavior discussed above will be expanded to pursue development of a modeling scheme that better ensures EA quality through the requirement elicitation, analysis, and specification phases. The communications path we propose will be further defined to include management and control techniques designed to handle EA design complexities and risk aspects associated with project scope and stakeholder requirement ambiguity.

The elicitation and documentation of each stakeholder desired requirement will include a provision for verification and validation via an agreed-upon, in writing, mechanism between stakeholders and the EIA. As with any scheme used by the EIA, the modeling of complex requirements must include a mechanism that provides full traceability for each requirement through transformation to a technical specification to final product deliverable. Therefore, we plan to explore establishing a process, a baseline, to handle requirement traceability from which future comparisons can be made between the current state of the requirement and its final technical specification.

References

[1] The DoD Architecture Framework Version 2.0 DoD Deputy Chief Information Officer, Department of Defense (April 2011)

[2] Chief Information Officers Council, Federal Enterprise Architecture Framework, CIO Council, Version 1.1 (August 5, 1999)

[3] Bakan, I., Tasliyan, M., Eraslan, I.H., Coskun, M.: The Effect of Technology on Organizational Behavior and the Nature of Word. In: IAMOT Conference, Washington, D.C. (2004)

[4] Beer, M.: Organizational Behavior and Development. Harvard Business Review, Harvard University

[5] Boudreau, M.-C., Robey, D.: Enacting Integrated Information Technology: A Human Agency Perspective. Organization Science 16(1), 3–18 (2005)

[6] Chung, L., Nixon, B.A., Yu, E., Mylopoulous, J.: Non-Functional Requirements in Software Engineering. Kluwer Academic Publishers, Boston

[7] Davidson, E.: Technological Frames perspective on Information Technology and Organizational Change. The Journal of Applied Behavioral Science 42, 23 (2006)

[8] Edwards, C.: Is Lack of Enterprise Architecture Partly to Blame for the Finance Industry Collapsing So Spectacularly Version 0.01 (February 29, 2009), `http://www.AgileEA.com`

[9] Ferreira, C., Cohen, J.: Agile Systems Development and Stakeholder Satisfaction: A South African Empirical Study. In: Proceedings 2008 Conference of South African Institute Computer Scientists and Information Technologists on IT Research in Developing Countries, pp. 48–55 (2008)

[10] Fowler, M.: Patterns of Enterprise Application Architecture. Addison-Wesley, New York (2003)

[11] Gauld, R.: Public Sector Information System Failures: Lessons from a New Zealand Hospital Organization. Government Information Quarterly 24(1), 102–114 (2007)

[12] Giddens, A.: The Constitution of Society: Outline of the Theory of Structuration. University of California Press (1984)

[13] Hanza, H.M.: Separation of Concerns for Evolving Systems: A Stability-Driven Approach. In: Workshop on Modeling and Analysis of Concerns in Software, MACS 2005, St. Louis, MO (May 2005)

[14] Herzberg, F., Mausner, B., Bloch Synderman, B.: The Motivation to Work. Wiley Johns & Sons, Inc. (January 1959)

[15] Kaur, R., Sengupta, J.: Software Process Models and Analysis on Failure of Software development Projects. Internation. Journal of Scientific & Engineering Research 2(2) (February 2011)
[16] Kepner, C., Tregoe, B.: The New Rational Manager. Princeton Research Press (1997)
[17] Lankhorst, M., von Drunen, H.: Enterprise Architecture Development and Modeling, Via Nova Architecture (March 2007)
[18] Lankhorst, M., et al.: Enterprise Architecture at Work:Modeling, Communication and Analysis, 2nd edn. Springer, Heidelberg (2009)
[19] Lawhorn, B.: More Software Project Failure, CAI (March 31, 2010)
[20] Lewin, R., Regine, B.: Enterprise Architecture, People, Process, Business, Technology, Institute for Enterprise Architecture Developments (Online), `http://www.enterprise-architecture.info/Images/Extended%20Enterprise/Extended%20Enterprise%20Architecture3.html`
[21] Lynne Markus, M., Robey, D.: Information Technology and Organizational Change: Causal Research in Theory and Research. Management Science 34(5) (May 1988)
[22] Maslov, A.: Motivation and Personality. Harper Collins (1987)
[23] McGregor, D.: The Human Side of Enterprise. McGraw-Hill (1960)
[24] Mezzanotte Sr., D.M., Dehlinger, J., Chakraborty, S.: Applying the Theory of Structuration to Enterprise Architecture Design. In: IEEE/WorldComp 2011, SERP 2011 (July 2011)
[25] Mezzanotte Sr., D.M., Dehlinger, J., Chakraborty, S.: On Applying the Theory f Structuration in Enterprise Architecture Design. In: IEEE/ACIS (August 2010)
[26] Minoli, D.: Enterprise Architecture A to Z. CRC Press, New York (2008)
[27] Orlikowski, W.: The Duality of Technology: Rethinking the Concept of Technology in Organization. Organization Science 3(3), 398–427 (1992)
[28] Pressman, R.S.: Software Engineering: A Practioner's Approach, 7th edn. McGraw-Hill Series in Computer Science, New York, NY (2010)
[29] The Open Group, TOGAF Version 9, The Open Group (2009)
[30] Roeleven, S., Sven, Broer, J.: Why Two Thirds of Enterprise Architecture Projects Fail, ARIS Expert Paper (Online) , `http://www.ids-scheer.com/set/6473/EA_-_Roeleven_Broer_-_Enterprise_Architecture_Projects_Fail_-_AEP_en.pdf`
[31] Scacchi, W.: Process Models in Software Engineering, Final Version in Encyclopedia of Software Engineering, 2nd edn. John Wiley and Sons, Inc., New York (2001)
[32] Sessions, R.: A Comparison of the Top Four Enterprise-Architecture Methodologies. MSDN Library (May 2007)
[33] Sommerville, I.: Software Engineering, 8th edn. Addison-Wesley Publishers, Harlow (2007)
[34] Summerville, I., Sawyer, P.: Requirements engineering: A Good Practice Guide. John Wiley & Sons, Ltd., Baffins Lane (2000)
[35] Yu, E., Strohmaier, M., Deng, X.: Exploring International Modeling and Analysis for Enterprise Architecture, University of Toronto, Information Sciences
[36] Zachman, J.: Concepts of the Framework for Enterprise Architecture, Information Engineering Services, Pty, Ltd.

A Middleware for Reconfigurable Distributed Real-Time Embedded Systems

Fatma Krichen, Bechir Zalila, Mohamed Jmaiel, and Brahim Hamid

Abstract. Managing reconfigurable Distributed Real-time Embedded (DRE) systems is a tedious task due to the substantially increasing complexity of these systems and the difficulty to preserve their real-time aspect. In order to resolve this increasing complexity, we propose to develop a new middleware, called RCES4RTES (Reconfigurable Computing Execution Support for Real-Time Embedded Systems), allowing the dynamic reconfiguration of component-based DRE systems. This middleware provides a set of functions ensuring dynamic reconfiguration as well as monitoring and coherence of such systems using a small memory footprint and respecting real-time constraints.

Keywords: Software architecture, dynamic reconfiguration, middleware, distributed real-time embedded systems.

1 Introduction

Dynamic reconfiguration consists in evolving the system from its current configuration to another configuration at runtime. A system configuration is defined as an architecture of software components. Such system can evolve by either architectural

Fatma Krichen
ReDCAD, University of Sfax, Tunisia
e-mail: fatma.krichen@redcad.org
IRIT, University of Toulouse, France
e-mail: fatma.krichen@irit.fr

Bechir Zalila · Mohamed Jmaiel
ReDCAD, University of Sfax, Tunisia
e-mail: bechir.zalila@enis.rnu.tn, mohamed.jmaiel@enis.rnu.tn

Brahim Hamid
IRIT, University of Toulouse, France
e-mail: brahim.hamid@irit.fr

R. Lee (Ed.): Software Engineering Research, Management and Appl. 2012, SCI 430, pp. 81–96.
springerlink.com

or behavioral reconfigurations. Architectural reconfigurations consist in modifying the system structure such as adding or removing components or connections. Behavioral reconfigurations consist in modifying the system behavior by updating, for example, the non-functional properties or the implementations of some components. The dynamic reconfiguration of DRE systems is more crucial since it requires more time and additional resources which are limited in these systems. In fact, such reconfiguration should preserve the real-time aspect (e.g. meeting of thread deadlines) and maintain the embedded character (e.g. small memory size).

To ensure the reliability and the coherence of reconfigurations, we consider the dynamic reconfiguration in the whole development process of embedded systems. *First*, we proposed, in a previous work, a model-based approach [7, 8] for specifying reconfigurable systems at a high level. To do so, we defined a new meta-model as a complement of the MARTE profile [12] to describe software concepts of reconfigurable distributed real-time embedded systems. We also proposed a UML profile as an implementation of this meta-model. *Second*, we performed, in a previous work, the verification of non-functional properties of such systems at design time. We verified temporal properties (e.g. meeting of thread deadlines, livelock freedom) and resource-related properties (e.g. CPU and memory usage) using the RMS scheduling algorithm and the Cheddar framework. *Third*, we are developing an MDA-based approach for reconfigurable distributed real-time embedded systems. The system models will be translated to platform specific models. Then, code will be automatically generated from these platform specific models. In order to preserve the correctness of a system, the generated code requires a middleware providing a set of functions to ensure dynamic reconfiguration at runtime as well as monitoring and coherence. This paper focuses on the description of this middleware.

The state-of-the-art in addressing the development of middleware supporting reconfigurable embedded systems has several limitations. *First*, to the best of our knowledge, there are no middleware which implement the mechanisms of dynamic reconfiguration as well as the reflexivity and the coherence of component-based DRE systems. The majority of existing middleware [15, 3] ensures the dynamic reconfiguration but not the reflexivity and the coherence of such systems. Therefore, several systems become incoherent after a reconfiguration without the possibility to know it. *Second*, some of the proposed middleware [6, 3, 15] do not support real-time systems and have an important memory footprint. *Third*, most of middleware ensuring the dynamic reconfiguration handle a limited set of reconfigurations, either architectural [6, 3] or behavioral [2]. There are no middleware which handle both architectural and behavioral reconfigurations.

In this work, we propose a middleware, called RCES4RTES, which supports reconfigurable DRE systems. Our middleware allows to perform dynamic reconfiguration using a small memory footprint. It supports both architectural and behavioral reconfigurations. It also provides a set of functions ensuring the monitoring and the coherence of reconfigurable systems and preserving the real-time constraints.

The remainder of this paper is organized as follows. In Section 2, we briefly present existing middleware handling distributed real-time embedded systems and particularly reconfigurable ones. Section 3 describes the main functions of the proposed middleware. Section 4 details the implementation of these functions. Then, in Section 5, we consider a case study that has dynamic reconfiguration requirements: a GPS (Global Positioning System). Finally, Section 6 concludes this paper and presents some future work.

2 Related Work

Some middleware have been proposed to support distributed real-time embedded systems and particularly reconfigurable ones. In the following, we present the most important of ones.

DynamicTAO [6] presents an extension of the TAO middleware [13] to support adaptive applications running on dynamic environments. It supports safe dynamic reconfiguration of scalable, high-performance and distributed systems. This middleware implements the mechanisms of concurrency, security and monitoring and the dynamic reconfigurations at runtime. Possible reconfigurations call migration, loading/unloading of components, etc. In order to simplify the dynamic reconfiguration process, the components of the system are grouped in libraries which are loaded and used at runtime. However, DynamicTAO has some limitations. It does not manage the system state during a reconfiguration and it does not support behavioral reconfigurations. Moreover, DynamicTAO is not well adapted for hard real-time embedded systems.

The Fault-tolerant Loadaware and Adaptive middlewaRe (FLARe) [3] extends TAO and supports DRE systems. It presents an efficient QoS-aware component middleware. This middleware allows to manage the fault tolerance and the recovery requirements of soft real-time systems. FLARe ensures dynamic reconfiguration according to resource availability. However, FLARe does not support hard real-time systems and handles only periodic threads.

The Component Integrated ACE ORB (CIAO) [15] presents a free implementation of the model component LwCCM [11] and the specification of Real-Time CORBA [10]. CIAO supports component-based real-time embedded systems. It provides mechanisms for the specification, the implementation and the deployment of components. It uses the aspect-oriented programming to support the separation and the composition of real-time aspects and configuration concerns. However, neither the nature nor the implementation of reconfigurations are specified. Moreover, CIAO does not support aperiodic threads scheduling.

SwapCIAO [2] is an extension of CIAO middleware supporting reconfigurable DRE systems. It dynamically updates the implementations of components using extensions provided by the LwCCM component model. In fact, SwapCIAO presents an efficient QoS-aware component middleware which allows to transparently update component implementations. However, it is very difficult to maintain and no concrete implementation for this middleware has been proposed.

PolyORB_HI [17] is inspired from the PolyORB middleware architecture [16]. It uses the schizophrenic architecture and its canonical services to ensure the communication among heterogeneous platforms. This middleware respects the restrictions of the Ravenscar Profile [5, 9] which forbids the use of certain structures of programming languages (Ada, C, etc.) to become statically analyzable. PolyORB_HI is composed of a minimal middleware core and several automatically generated services. The minimal core presents the common services for all applications while the generated functions are customizable to the needs of the target application. Thanks to its very small memory footprint, PolyORB_HI presents an ideal middleware for embedded systems. However, it supports neither dynamic reconfiguration nor reflexivity.

In Table 1, we summarize the most important properties of the previously represented middleware. Given the comparative study results, our middleware handles the dynamic reconfiguration as well as the monitoring and the coherence like DynamicTAO and SwapCIAO. However, these middleware support object-oriented soft real-time systems while the proposed middleware handles the component-based hard real-time embedded systems like PolyORB_HI and CIAO. But, PolyORB_HI and CIAO middleware do not ensure the monitoring and the coherence of these systems. CIAO supports the dynamic reconfiguration contrary to PolyORB_HI, but it has a big memory footprint ($\simeq$ 5 MB). PolyORB_HI has a smaller memory footprint ($\simeq$ 70 KB). Contrary to the presented middleware, our middleware handles both architectural and behavioral reconfigurations using a small memory footprint.

Table 1 Middleware for distributed real-time embedded systems

	Middleware					
characteristics	DynamicTAO	CIAO	PolyORB_HI	FLARe	SwapCIAO	RCES4RTES
Granularity	Object (RT-CORBA)	Component (LwCCM)	Component	Object (RT-CORBA)	Object (LwCCM)	Component
Hard real-time	no	yes	yes	no	N/A	yes
Memory size	1,5 MB	5 MB	70 KB	10 MB	N/A	131 KB
Monitoring	yes	N/A	no	yes	yes	yes
Dynamic reconfiguration	yes	yes	no	yes	yes	yes
Portability	yes (C++)	yes (C++)	yes (Java)	yes (C++)	yes (C++)	yes (java)
Coherence	yes (difficult)	N/A	no	no	yes	yes

3 Proposed RCES4RTES Middleware

Our RCES4RTES middleware supports DRE systems. It defines a distributed system by a set of interconnected nodes. Only the nodes which exchange data should be connected to reduce resource usage. Each node is identified by its name and supports a set of components that should be deployed on it.

A component is characterized by non-functional properties and represented as one of the supported threads:

- Periodic thread has a constant time interval between two executions and it is defined by three parameters: deadline, period and execution requirement.
- Sporadic thread can be executed at arbitrary times with defined minimum inter-arrival time between two consecutive executions. It is also identified by a deadline and an execution requirement.
- Aperiodic thread is activated only once and it is characterized by an arrival time and an execution requirement.
- Hybrid thread is the combination of both periodic and sporadic thread characteristics. It requires another thread to notify it by the arrival of its period using events.

The components are connected using ports classified into output or input ports. Our middleware supports two kinds of connectors: delegation connector (i.e. between two ports of the same kind) and assembly connector (i.e. between an output port and an input port). The connections between components are identified by assigning the destination ports of each port.

As shown in Figure 1, the central function of the proposed middleware is the dynamic reconfiguration of software component-based DRE systems. This RCES4RTES middleware also provides other functions required for developing reconfigurable real-time embedded applications. Our middleware consists in:

- Supporting the monitoring of the system by supervising at runtime the topology and the behavior of the architecture, tracing the system execution (e.g., getting the number of components and connections) and updating the shared variables. The monitoring function can also be used to assure the reflexivity of the system.
- Preserving the coherence of the system during and after reconfigurations since reconfiguration may lead the system to incoherent states.
- Respecting real-time constraints. In fact, on each node of the system, a *dynamic reconfiguration thread* is automatically created. It represents a sporadic thread applying reconfiguration actions. It will be considered as a system thread and then is scheduled with the other system threads. Using this sporadic thread, our middleware can easily manage the reconfigurations without affecting the system threads execution and exceeding thread deadlines.
- Ensuring the communication among heterogeneous platforms using the schizophrenic architecture and its canonical services. Taking advantage of these services and contrary to the existing middleware, the RCES4RTES middleware has a small memory footprint ($\simeq$ 131 KB).
- Respecting the restrictions of Ravenscar profile [5] to ensure the schedulability, the deadlock freedom and the livelock freedom of system threads.

Further details about these functions and their advantages are described in the following sub-sections.

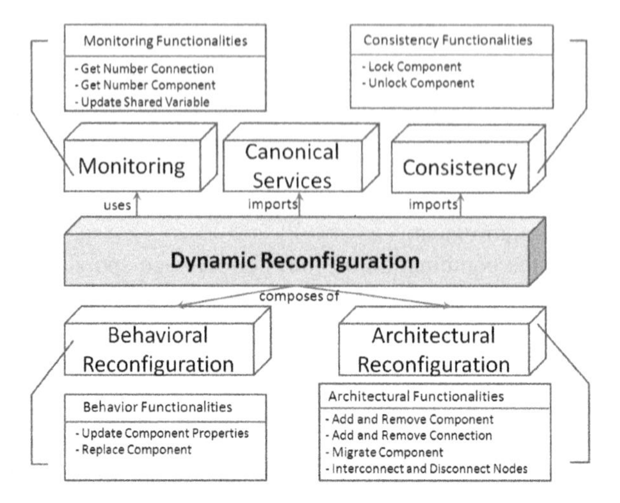

Fig. 1 RCES4RTES middleware

3.1 Dynamic Reconfiguration

Our RCES4RTES middleware performs the dynamic reconfiguration of DRE systems through two types of reconfiguration: (i) the architectural reconfiguration for modifying the software architecture such as adding or removing components or connections, (ii) the behavioral reconfiguration for modifying the components behavior by updating, for example, the component properties.

Our middleware handles the following architectural reconfigurations:

- *Connect nodes*: adding a communication channel between two nodes,
- *Disconnect nodes*: deleting a communication channel between two nodes,
- *Add connection*: adding a communication channel between two components,
- *Remove connection*: deleting a communication channel between two components,
- *Add component*: creating and deploying component on execution platform and then connecting it with the other components,
- *Remove component*: removing its connections with the other components, undeploying it and then deleting it,
- *Migrate component*: moving a component from a node to another

Our middleware also handles the following behavioral reconfigurations:

- *Update component properties*: modifying the component properties by affecting them with new values,
- *Replace component*: updating the implementation of a component.

Our RCES4RTES middleware should thus introduce a set of data structures (i.e. hash tables) to ensure the previous reconfigurations and to update the current state of the application in terms of nodes, components, connections and ports. First, an

IntNodesDS data structure is defined in each node to describe the connections between the current node and the other nodes and to manage the dynamic interconnection of nodes. Each pair of nodes having at least one connection between their associated components should be connected. Connecting and disconnecting nodes require the update of the corresponding data structure. Second, our middleware introduces a *destinationPortsDS* data structure to update at runtime the state of the interconnection between components. In each node, this data structure presents the destination ports of each deployed component port. Adding or removing connector between two ports (i.e. components) requires updating the corresponding data structure. Third, two other data structures for each node are defined (*ComponentsDS* and *portsDS*) for adding, removing and migrating components at runtime. The *ComponentsDS* data structure contains all application nodes with their corresponding deployed component instances while the *portsDS* data structure contains all application component instances with their related ports. When a component have been added, removed or migrated, these two data structures are updated in each node to ensure the system coherence.

A set of routines is provided in our middleware implementing the previously presented architectural and behavioral reconfigurations. For example, Listing 1 presents the routine which allows to add connection between two components (i.e. two ports). First, this routine verifies whether these two components are nested in which case both ports *P1* and *P2* must have the same kind, i.e. input or output port (lines 2–4). Otherwise, the port *P1* must be an output port and the port *P2* must be an input port (lines 5 and 6). To avoid any data loss between components, we lock the source component (i.e. connected by its port *P1*) before adding the connection (line 8). Just after the creation of connection, the source component must be unlocked (line 10) to send and receive data. The connection will be created by adding the port *P2* identifier to the set of destinations of the port *P1* (line 9) (i.e. updating the corresponding data structure *destinationPortsDS*).

```
1  addConnection(P1,P2:Port)
2     if ((isNested(P1.getComponent(),P2.getComponent())
3           and ((isOutputPort(P1) and isOutputPort(P2))
4                 or (isInputPort(P1) and isInputPort(P2))))
5                 or
6        (not isNested(P1.getComponent(),P2.getComponent())
7           and isOutputPort(P1) and
8           isInputPort(P2))
9           and
10       not isExistConnection(P1,P2)) then
11       lockComponent(P1.getComponent());
12       P1.addDestinationPort(P2);
13       unlockComponent(P1.getComponent());
14    endIf
```

Listing 1. Adding connection

Listing 2 presents the routine which allows to add a component. This routine, called `addComponent`, is used for creating a new component instance to a node and connecting it to other components. It has as parameters (line 1): the component, the node where the new instance will be deployed, the destination list of each port of this new instance and the list of ports which have a port of this instance in their destinations. After verifying whether the node *n* supports this component, a new instance of component *t* will be created and deployed in this node (lines 2–4). The connections of this instance with the other components will be added (as shown in lines 6–15).

```
1  addComponent(t: ComponentType; n:Node; dest[][],
2                              source[][]:hash table)
3     if (exist(t,n)) then
4        c=newInstance(t);
5        deploy(c,n);
6        For i=1 to dest.length do
7           P1=getOutputPort(dest[i][1]);
8           P2=getInputPort(dest[i][2]);
9           addConnection(P1,P2);
10       EndFor
11       For i=1 to source.length do
12          P1=getOutputPort(source[i][2]);
13          P2=getInputPort(source[i][1]);
14          addConnection(P2,P1);
15       EndFor
16    endIf
```

Listing 2. Adding component

As an example of behavioral reconfiguration, Listing 3 presents the routine designed to update component properties. The set of properties to be updated is first classified into critical and non-critical properties (line 2). The non critical properties can be updated at runtime without locking the corresponding component (lines 3–5) while the critical properties which affect the component execution will be updated (lines 8–10) after locking the corresponding component (line 7).

```
1  updateProperties(c: ComponentInstance; P[]:properties)
2     P.classifyProperties(criticalProperties,
3                          nonCriticalProperties);
4     For i=1 to nonCriticalProperties.length do
5        updateProperty(nonCriticalProperties[i],c);
6     EndFor
7     if (isExist(criticalProperties)) then
8        lockComponent(c);
9        For i=1 to criticalProperties.length do
10          updateProperty(criticalProperties[i],c);
11       EndFor
12       unlockComponent(c);
13    endIf
```

Listing 3. update of component properties

3.2 *Coherence*

The coherence is an essential property of a reconfigurable system. The system should be in a correct state during and after reconfigurations to prevent failures. To maintain the correct state of a system, we should avoid message loss between components during reconfigurations. Each component affected by the reconfiguration process should be locked during the reconfiguration. For this, we define two routines for locking and unlocking components. The locking of component consists in preventing the source components (i.e. the components which send requests to the locked component) to send requests and achieving the treatment of all current requests. The unlocking of component consists in releasing the lock by allowing the source components to send requests.

To avoid wasting time and to minimize as possible the locking duration, locking and unlocking components should be made respectively after creating new components and before deleting components in the reconfiguration routines.

3.3 Monitoring

Our middleware provides routines for monitoring DRE systems. Using these routines, we can observe a system state during its execution. The monitoring of such a system at runtime helps us manage the dynamic reconfiguration. For this, routines are implemented for:

- getting the component number running on the system,
- getting the connection number between a given component and other components,
- getting the last read/write access time to shared variables.

3.4 Hard Real-Time

The time of reconfiguration T_{rd} (presented by the Equation 1) is the sum of the locking duration of components T_b, the execution time of reconfiguration actions T_{act} and the transfer time of component state T_{state}. T_{state} is defined only in the case of the migration of component from a node to another while T_{act} is the execution time of the dynamic reconfiguration thread. For each node, a sporadic thread (i.e dynamic reconfiguration thread) is created to perform the reconfigurations on this node and to inform the other nodes of these reconfigurations. This thread allows respecting time constraints (i.e. all system threads should meet their deadlines during and after reconfigurations). It will be considered and scheduled with system threads.

$$T_{rd} = T_b + T_{state} + T_{act} \tag{1}$$

T_b, T_{state} and T_{act} are proportional to the size of data to be treated. T_{act} also depends on the processor frequency and the transport layer. To get a deterministic reconfiguration time, we can so use a deterministic transport layer such as SpaceWire [1].

3.5 Conformance to the Ravenscar Profile

The Ravenscar profile [5, 9] introduces restrictions allowing the schedulability and the deadlock and livelock freedom for real-time embedded systems. For efficiently performing the scheduling, the Ravenscar profile avoids the use of threads which are randomly launched. It requires periodic and sporadic threads. Therefore, the set of threads to be analyzed is fixed and has static properties. Ravenscar profile also requires asynchronous communications. The communications between threads should be ensured only by a static set of protected shared objects and the access to these protected objects should be done using PCP (Priority Ceiling Protocol) [14].

In our approach, we are interested in four kinds of threads: periodic, sporadic, aperiodic and hybrid threads. As each aperiodic thread has an arrival time and each hybrid thread presents the combination of both periodic and sporadic thread characteristics, these threads (aperiodic and sporadic threads) respect the Ravenscar profile restrictions.

We also use asynchronous communications between system threads using PCP protocol. As we are focused on distributed systems, the time of both construction and sending of messages is non-deterministic because of the non-reliability of transport layers. This can be a source of message loss. To resolve this limitation, we propose to use a reliable and real-time transport layer such as SpaceWire [1]. We can therefore consider our distributed system as a local system.

4 Implementation of RCES4RTES Middleware

In this section, we briefly describe the implementation of the proposed middleware and particularly focus on the implemented routines ensuring the dynamic reconfiguration. For efficiently developing the proposed middleware functions, the RTSJ (i.e. Real Time Specification for Java) [4] is used. It provides adapted routines to overcome the drawbacks of Java for real-time systems. Moreover, RTSJ allows the real-time threads scheduling and the deterministic memory management by introducing new memory areas and solving the problem of garbage collecting in Java.

For developing our RCES4RTES middleware, we updated and extended the PolyORB_HI middleware implementation with the previously mentioned functions in section 3. PolyORB_HI is selected in consideration of the following advantages: (i) it has a very small memory footprint, (ii) it is portable, (iii) it supports component-based hard real-time embedded systems and (iv) it supports three thread kinds: periodic, sporadic and hybrid threads.

Indeed, preexisting routines are updated and new routines are added to support reconfigurable distributed real-time embedded systems. For this purpose, we first adapted the implementation of canonical services such as the addressing service which manages the components references at runtime. Hash tables are used for adding and removing threads (i.e. components) to nodes at runtime.

Contrary to PolyORB_HI which connects all defined application nodes, our middleware connects only the nodes which exchange data. This solution reduces resource usage by connecting and disconnecting nodes at runtime. For this, we modified the PolyORB_HI *connect* routine to connect nodes at runtime and then we added a new routine, called *disconnect*, to disconnect nodes at runtime. Other new routines designed for the dynamic reconfiguration have been also developed:

- *addTask* allows to add a new thread to a node. Each thread is defined by its type (periodic, sporadic, hybrid or aperiodic) and a set of non-functional properties (period, deadline, etc.). Each new thread should be added to the corresponding hash tables. Then, the related connections to other system threads should be created through adding the destination ports to the corresponding hash tables.
- *removeTask* allows to remove a thread from a node. All connections of this thread should thus be deleted by removing the destination ports from the corresponding hash tables. Then, this thread should be removed from the corresponding hash tables.
- *migrateTask* allows to migrate a thread from a node to another. The state of this thread should be stored in a table which has been sent from the source node to the target node. A new thread that takes the stored state in this table should thus be created on the target node and the first thread in the source node should be deleted.
- *changeProperties* allows to modify the properties of thread. The non-critical properties are changed directly while the critical ones are changed after locking the associated thread.

5 A GPS Case Study

In this section, we consider a well-known example with dynamic reconfiguration requirements: a GPS (Global Positioning System) [8]. In fact, a GPS is a radio navigation system which provides accurate navigation signals to any place on the earth. The GPS system consists of three nodes (Figure 2): *GPS_Terminal*, *GPS_Satellite* and *GPS_ControlBase*. *GPS_ControlBase* receives and sends information to *GPS_Satellite* in order to synchronize the satellite clocks. *GPS_Satellite* sends to *GPS_Terminal* an encrypted signal containing various information useful for localization and synchronization.

For the sake of simplicity, many functionalities of this case study have been omitted. Both satellite and control base are represented by basic components (resp. *GpsSatellite* and *GpsControlBase*). In this paper, we only describe the *GPS_Terminal* architecture which consists of five components:

- *Position* component for receiving the satellite signal,
- *Receiver* component for converting the analog signal into a digital signal,
- *Decoder* component for decoding digital information and separating between the information to calculate distance and time information,

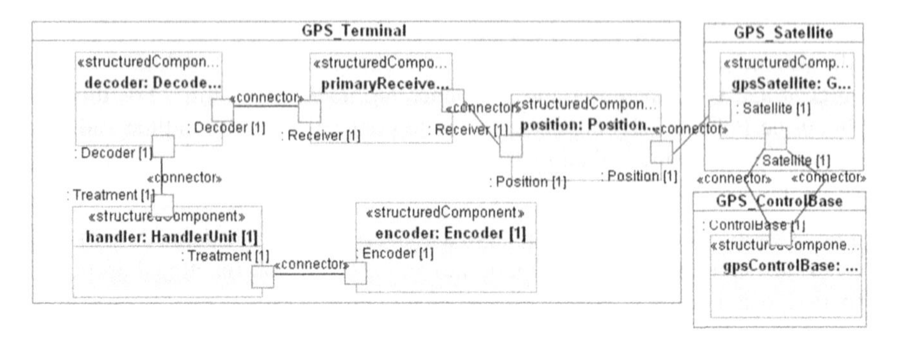

Fig. 2 Insecure GPS Configuration

- *HandlerUnit* component for computing the distance from the satellite in order to obtain the position,
- *Encoder* component for encoding time and position information.

These components define the *GPS_Terminal* with insecure functioning which consists in a traditional (or public) use of a GPS. The non-functional properties of each component are given in Table 2. Due to the complexity of this case study, these properties values are obtained from simple simulations.

Table 2 Non-functional properties of GPS components

Component type	Nature	Period	Deadline	WCET
GpsSatellite	Periodic	400 ms	400 ms	30 ms
GpsControlBase	Sporadic	400 ms	400 ms	30 ms
Position	Sporadic	100 ms	100 ms	20 ms
SecurePosition	Sporadic	100 ms	100 ms	20 ms
AccessController	Sporadic	100 ms	100 ms	20 ms
Receiver	Sporadic	100 ms	100 ms	20 ms
Decoder	Sporadic	100 ms	100 ms	20 ms
HandlerUnit	Sporadic	100 ms	100 ms	20 ms
Encoder	Sporadic	100 ms	100 ms	20 ms

To illustrate the use of our middleware, we define two configurations of GPS: *Insecure GPS Configuration* (Figure 2) as initial configuration and *Secure GPS Configuration* (Figure 3). The *Secure GPS Configuration* presents the *GPS_Terminal* with secure functioning. It describes a restricted use of a GPS with some safety requirements.

Figure 4 shows an execution trace of the GPS terminal in the *Insecure GPS Configuration*. The terminal starts its execution by the creation of component instances such as the creation of a *position* instance of *Position* component. Then, the terminal receives information from satellite to establish the road to be followed.

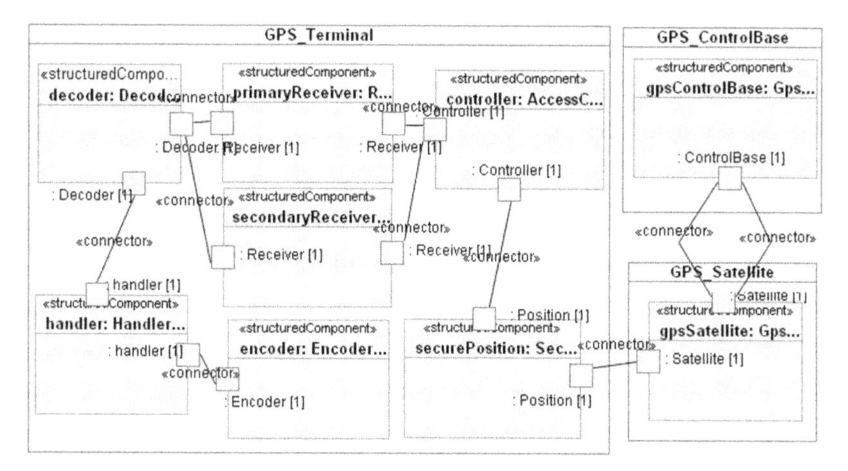

Fig. 3 Secure GPS Configuration

```
Creation of the position instance on GPS_Terminal node
position : wait initialization
position : new Dispatch
position is waiting for incoming events
Creation of the primaryReceiver instance on GPS_Terminal node
primaryReceiver : wait initialization
primaryReceiver : new Dispatch
primaryReceiver is waiting for incoming events
Creation of the decoder instance on GPS_Terminal node
decoder : wait initialization
decoder : new Dispatch
decoder is waiting for incoming events
Creation of the treatmentUnit instance on GPS_Terminal node
HandlerUnit : wait initialization
Creation of the encoder instance on GPS_Terminal node
encoder : wait initialization
encoder : new Dispatch
HandlerUnit : new Dispatch
encoder is waiting for incoming events
HandlerUnit is waiting for incoming events
GPS_Terminal node : receives 18 bytes from GPS_Satellite node
position : store received message
```

Fig. 4 Trace of the execution of the GPS terminal in the *Insecure GPS Configuration*

Our middleware allows the dynamic reconfiguration by switching the GPS from *Insecure GPS Configuration* to *Secure GPS Configuration*. This switching consists in removing all instances of Position component and adding instances of both *SecurePosition* and *AccessController* components to assure the secure reception and the satellite signal control. For this, the connections of *position* instance with *gpsSatellite* and *primaryReceiver* instances should be removed and two instances of both *SecurePosition* and *AccessController* with their connections should be added. For the *Secure GPS Configuration*, a second instance of *Receiver* component should also be added. The following reconfigurations should be sequentially handled to

```
Dynamic reconfiguration: the securePosition whose type is SecurePosition has been added
on GPS_Terminal node
Dynamic reconfiguration: the controller whose type is AccessController has been added
on GPS_Terminal node
Dynamic reconfiguration: the secondaryReceiver whose type is Receiver has been added
on GPS_Terminal node
GPS_Terminal node: send 18 bytes to GPS_Satellite node
Dynamic reconfiguration: distant connection between the port 17 and 11 has been added
Dynamic reconfiguration: local connection between the port 10 and 13 has been added
Dynamic reconfiguration: local connection between the port 12 and 3 has been added
Dynamic reconfiguration: local connection between the port 12 and 15 has been added
Dynamic reconfiguration: local connection between the port 14 and 5 has been added
Creation of the securePosition instance on GPS_Terminal node
securePosition : wait initialization
securePosition : new Dispatch
securePosition is waiting for incoming events
Creation of the controller instance on GPS_Terminal node
Creation of the secondaryReceiver instance GPS_Terminal node
Coherence: the position has been locked
controller  : wait initialization
Consistency: the position instance has been unlocked
secondaryReceiver : wait initialization
Dynamic reconfiguration: local connection between the port 0 and 3 has been removed
controller  : new Dispatch
secondaryReceiver : new Dispatch
isconnection : existe=true
controller  is waiting for incoming events
secondaryReceiver is waiting for incoming events
GPS_Terminal node : send 18 bytes to GPS_Satellite node
Dynamic reconfiguration: distant connection between the port 17 and 1 has been removed
Dynamic reconfiguration: position has been removed from the GPS_Terminal node
```

Fig. 5 Part of trace of the dynamic reconfiguration from *Insecure GPS Configuration* to *Secure GPS Configuration*

guarantee the switching from *Insecure GPS Configuration* to *Secure GPS Configuration* (as shown in Figure 5):

- Adding *controller* instance of *AcsessController* component,
- Adding *securePosition* instance of *SecurePosition* component,
- Adding *secondaryReceiver* instance of *Receiver* component,
- Adding connection between *gpsSatellite* and *securePosition*,
- Adding connection between *securePosition* and *controller*,
- Adding connection between *controller* and *primaryReceiver*,
- Adding connection between *controller* and *secondaryReceiver*,
- Adding connection between *secondaryReceiver* and *decoder*,
- Removing connection between *position* and *primaryReceiver*,
- Removing connection between *gpsSatellite* and *position*,
- Removing *position* instance of *Position* component.

After applying efficiently the previously presented reconfiguration actions, we demonstrate also that the monitoring and the coherence are ensured by our middleware using the considered GPS case study. After the transition from *Insecure GPS Configuration* to *Secure GPS Configuration*, the system remains coherent and preserves its temporal constraints. As shown in Figure 6, we observe that both *primaryReceiver* and *secondaryReceiver* instances run normally and meet their deadlines.

In addition to the small memory footprint of our middleware ($\simeq$ 131 KB), we have computed the memory footprint of the GPS case study on each system

```
secondaryReceiver calls a sub program at (1321967893590 ms, 445144 ns)
primaryReceiver, getValue of input port 3
primaryReceiver, readIn : reading the oldest element in the queue of event [data]
primaryReceiver, readIn : value read from input port 3
primaryReceiver, readIn : reading the oldest element in the queue of event [data]
primaryReceiver, readIn : value read from input port 3
primaryReceiver, getValue done
primaryReceiver, nextValue for event [data] input port 3
primaryReceiver, dequeue : dequeuing event [data] input port 3
primaryReceiver, historyIncrementFirst : globalHistoryFirst = 4
primaryReceiver calls a sub program at (1321967893642 ms, 430784 ns)
secondaryReceiver termine impl at (1321967893647 ms, 293352 ns)
 at t=14h18m13s647ms ***** the node 0 converted the signal into the value : 15
secondaryReceiver, storeOut : storing value for output port 14
secondaryReceiver, storeOut : value stored for output port 14
secondaryReceiver, sendOutput for output port 14
secondaryReceiver, setInvalid : setting invalid for output port 14
secondaryReceiver, sendOutput output port 14
secondaryReceiver, send output to input port 5 of entity 2
decoder : store received message
.....
decoder calls a sub program at (1321967893648 ms, 987482 ns)
primaryReceiver termine impl at (1321967893716 ms, 950898 ns)
```

Fig. 6 Meeting of thread deadlines

node: 51.5 KB for the *GPS_Terminal* node, 25.6 KB for the *GPS_Satellite* node and 25.9 KB for the *GPS_ControlBase* node. So, we can conclude that the memory footprint for each node is small.

6 Conclusion and Future Work

In this paper, we developed a middleware, called RCES4RTES (Reconfigurable Computing Execution Support for Real Time Embedded Systems), that manages the dynamic reconfiguration of component-based DRE systems. This middleware ensures the monitoring and the coherence. It also allows preserving the real-time aspect and conserving the embedded character during and after reconfigurations. Our proposed middleware has been implemented using RTSJ and the existing PolyORB_HI middleware which has been updated and extended to the implemented solution. After that, a case study (a GPS system) has been considered to illustrate how the dynamic reconfiguration is performed and to verify that the monitoring and the coherence of such a reconfigurable system are ensured.

Our middleware manages the complexity of DRE systems and especially reconfigurable ones. It ensures the reconfiguration of the system at runtime without losing the coherence of the system. It also supports both behavioral and architectural reconfigurations using a smaller memory footprint ($\simeq$ 131 KB) compared to the existing middleware supporting the dynamic reconfiguration.

As future work, we plan to extend our middleware in order to support fault tolerance. We also suggest to handle the security aspects of components addition/removal in a malicious way.

Acknowledgements. The authors thank Alvine BOAYE BELLE for her valuable work when implementing the first release of the RCES4RTES.

References

1. SpaceWire - Links, nodes, routers and networks. European Space Agency. Technical report (2008)
2. Balasubramanian, J., Natarajan, B., Schmidt, D.C., Gokhale, A., Parsons, J., Deng, G.: Middleware Support for Dynamic Component Updating. In: Meersman, R. (ed.) OTM 2005, Part II. LNCS, vol. 3761, pp. 978–996. Springer, Heidelberg (2005)
3. Balasubramanian, J., Tambe, S., Lu, C., Gokhale, A.S., Gill, C.D., Schmidt, D.C.: Adaptive failover for real-time middleware with passive replication. In: Proceedings of the 15th IEEE Real-Time and Embedded Technology and Applications Symposium, pp. 118–127 (2009)
4. Bruno, E.J., Bollella, G.: The Real-Time Specification for Java. Prentice Hall (2009)
5. Burns, A.: The Ravenscar Profile. Ada Letter XIX, 49–52 (1999)
6. Kon, F., Román, M., Liu, P., Mao, J., Yamane, T., Magalhães, L.C., Cambell, R.H.: Monitoring, Security, and Dynamic Configuration with the *dynamicTAO* Reflective ORB. In: Coulson, G., Sventek, J. (eds.) Middleware 2000. LNCS, vol. 1795, pp. 121–143. Springer, Heidelberg (2000)
7. Krichen, F., Hamid, B., Zalila, B., Coulette, B.: Designing dynamic reconfiguration of distributed real time embedded systems. In: Proceedings of the 10th Annual International Conference on New Technologies of Distributed Systems, NOTERE. IEEE, Tozeur (2010)
8. Krichen, F., Hamid, B., Zalila, B., Jmaiel, M.: Towards a Model-Based Approach for Reconfigurable DRE Systems. In: Crnkovic, I., Gruhn, V., Book, M. (eds.) ECSA 2011. LNCS, vol. 6903, pp. 295–302. Springer, Heidelberg (2011)
9. Kwon, J., Wellings, A., King, S.: Ravenscar-Java: a high integrity profile for real-time Java. In: Proceedings of the joint ACM-ISCOPE Conference on Java Grande, pp. 131–140 (2002)
10. OMG: Real-time CORBA Specification (2005), `http://www.omg.org/technology/documents/specialized_corba.htm#RT_CORBA`
11. OMG: CORBA Specification, Version 3.1. Part 3: CORBA Component Model (2008), `http://www.omg.org/technology/documents/corba_spec_catalog.htm#CCM`
12. OMG: A UML Profile for MARTE: Modeling and Analysis of Real-Time Embedded systems, Beta 2 (2009), `http://www.omgmarte.org/Documents/Specifications/08-06-09.pdf`
13. Schmidt, D.C., Cleeland, C.: Applying a pattern language to develop extensible ORB middleware. Cambridge University Press (2001)
14. Sha, L., Rajkumar, R., Lehoczky, J.P.: Priority inheritance protocols: An approach to real-time synchronization. IEEE Transactions on Computers 39(9), 1175–1185 (1990)
15. Subramonian, V., Deng, G., Gill, C., Balasubramanian, J., Shen, L.J., Otte, W., Schmidt, D.C., Gokhale, A., Wang, N.: The design and performance of component middleware for qos-enabled deployment and configuration of dre systems. Journal of Systems and Software 80, 668–677 (2007)
16. Vergnaud, T., Hugues, J., Pautet, L., Kordon, F.: PolyORB: A Schizophrenic Middleware to Build Versatile Reliable Distributed Applications. In: Llamosí, A., Strohmeier, A. (eds.) Ada-Europe 2004. LNCS, vol. 3063, pp. 106–119. Springer, Heidelberg (2004)
17. Zalila, B., Pautet, L., Hugues, J.: Towards Automatic Middleware Generation. In: Proceedings of the International Symposium on Object-Oriented Real-time Distributed Computing, pp. 221–228. IEEE (2008)

Do We Need Another Textual Language for Feature Modeling?

A Preliminary Evaluation on the XML Based Approach

Jingang Zhou*, Dazhe Zhao, Li Xu, and Jiren Liu

Abstract. Feature Modeling (FM) is an essential activity for capturing commonality and variability in software product lines. Most of today's FM tools are graphical and represent feature models as feature diagrams (FDs). Though FDs are intuitive at first sight, they generally lack of expressiveness and vague in syntax. To overcome these problems, some textual languages are proposed with a richer expressiveness and formal semantics. But are these languages superior than existing modeling approach as they stated, e.g., XML-based one, which is standard based and has vast of acceptance of application other than FM. In this paper, we elaborate the XML-based textual feature modeling approach, evaluate it from multi perspectives, and compare it with another representative textual FM language —TVL, a recently published language. We demonstrate the advantages and disadvantages of the XML-based approach, and argue that the XML-based approach is still the most available and pragmatic approach for adoption in industry though it has some clear limitations which can be fed to more advanced tool support.

Keywords: feature modeling, XML, software product line, textual language.

Jingang Zhou
College of Information Science and Engineering, Northeastern University,
110004, Shenyang, China
e-mail: zhou-jg@computer.org

Dazhe Zhao · Li Xu · Jiren Liu
State Key Laboratory of Software Architecture (Neusoft Corporation),
110179, Shenyang, China
e-mail: {zhaodz,xuli,liujr}@neusoft.com

* Corresponding author.

R. Lee (Ed.): Software Engineering Research, Management and Appl. 2012, SCI 430, pp. 97–111.
springerlink.com

1 Introduction

Software product line (SPL) [1, 2] is an advanced and systematic software development paradigm in the past 20 years. In a SPL, a product is built from core assets (generally common requirements, architectures, components, frame-works, etc.) rather individually from scratch. The differences among product members are captured by their features. A feature is "a distinguishable characteristic of a concept (e.g., system, component, and so on) that is relevant to some stakeholder of the concept" [3]. With the abstraction of feature, many important characteristics of a SPL can be gained via its feature model. Therefore, feature modeling (FM) is an essential activity for software product line engineering (SPLE).

From its introduction in FODA [4] and following augments in subsequent research and practices, feature models are generally represented in feature diagrams (FDs), which display a feature model in a hierarchically arranged set of features graphically like a tree. This can be seen from most of today's FM tools [5], commercial (e.g., Pure::Systems[1]) or free (e.g., XFeature [2] and FeaturePlugin [6]). However, most of such tools (and feature models) suffer from scalability in industry settings [5, 7] (e.g., editing for large feature models with more than 10,000 features), no support for attributes concepts [8], and vague syntax of graphical nature [8]. To overcome these problems and also provide complementary view for the graphical ones, some textual dialects have been proposed, like FDI [9], GUIDSL [10], TVL [11, 8], and CML [12] (there are also some other textual approaches like FeaturePlugin, FAMA [13], and pure::variants, which are XML-based and only read or written by the tools not targeted at engineers). In these approaches, only TVL can be seen a true textual FM language which provides formal syntax and semantics and explicitly targets for textual FM with a reference implementation[3], while current CML is too immature to use (concepts only without implementation) and FDI and GUIDSL are only for feature analysis and product configuration purpose other than textual FM.

However, though many FM tools leverage XML only as FDs serialization or deserialization underneath the tool with some specific tags, it is still natural to image XML as an explicit textual FM language due to the tree-like grammar or nature which is same as to a feature model. But, the authors of TVL have not compared TVL with the XML-based approach[4]. This is mainly because, in our view, there is not an approach which explicitly targets for XML-based textual FM activity. So, it is time to uncover the power of XML in the context of textual FM environment.

---

[1] http://www.pure-systems.com/. It also has a free edition.
[2] http://www.pnp-software.com/XFeature/
[3] https://www.info.fundp.ac.be/tvl/
[4] To be noted, in this paper, the XML-based approach means not only specific to XML, but also general related standards like XML Schema, XSL, etc.

In this paper, we explore the potential of XML as a textual FM language in a tutorial way and argue that the XML-based approach is more pragmatic for the industry. We believe it is important that both researchers and practitioners (especially those in industry) reexamine the XML-based way, avoid reinventing the wheels in languages design, and benefit from a standard-based approach since XML has a vast acceptance in the industry.

In the remainder of the paper, we present some basic FM concepts in Section 2 with a typical FD upon which the XML-based textual FM approach is elaborated in Section 3. Evaluation and comparison to TVL are illustrated in Section 4 which shows both advantages and limitations of our approach with solutions for improvement. Threats to validity are addressed in Section 5. We conclude in Section 6.

2 Feature Modeling Concepts

A feature model is a hierarchically arranged set of features, which is used to model a SPL in terms of features and relations among them. Originating from FODA, there are many FD language methods [14, 15, 16, 17, 18] to augment FODA with more expressiveness. To make the representation general and neutral, we adapt[5] the *cardinality-based feature model* (CFM) [16] method since CFM has formal semantics and is the most comprehensible and flexible as well as being one of the most quoted [19, 20]. Fig.1 demonstrates the adapted CFM for a partial HIS SPL.

A feature model is composed by a **root** feature (*HIS* in Fig.1) and a set of **constraints** (divided into **global** constraints and **local** constraints. The former refers to constraints that involve features belonging to different feature group or parent (e.g., the constraint of "*Power line* **requires** *appliance control*"), while the latter refers to constraints among a single feature group or parent (e.g., the exclusive constraint among *Power line*, *ADSL*, and *Wireless*). A root is composed by some **solitary** features (e.g., *Control*, *Services*, and *light control*) which again are composed by some solitary features and **grouped** features (e.g., *Video on demand* and *Internet connection*). A solitary feature is qualified by its feature **cardinality** which is a sequence of intervals of the form $[n_1..n_1']...[n_l..n_l']$ indicating the allowed occurrence times in a SPL ([1..1] denotes the mandatory feature in FODA, e.g., *Control*; while [0..1] indicates an optional one, e.g., *appliance control*). A grouped feature (e.g., *ADSL*) must occur in a feature group (e.g., subfeatures of *Internet connection*, which form an alternative relation). We introduce the **attribute** feature (e.g., *PRICE*) notation from [20]. In the next section, we will map these feature model elements to XML representation.

---

[5] We used the representation of [20] for attribute feature and eliminate the representation of parameterized non-leaf features.

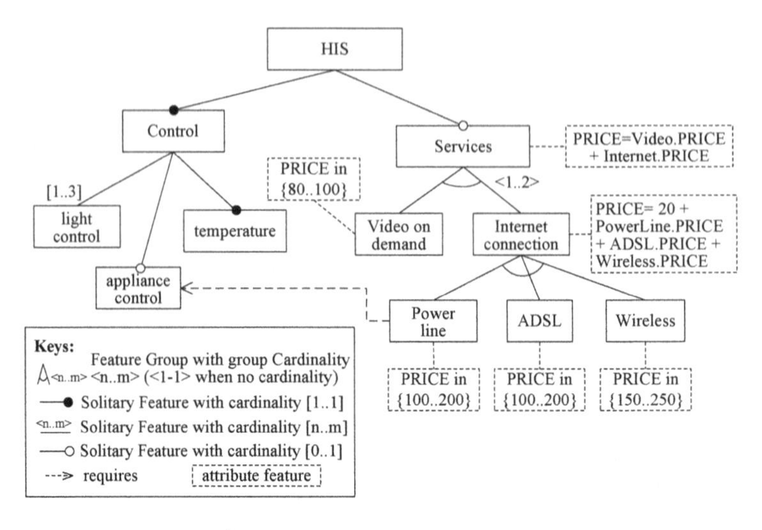

Fig. 1 Adapted feature model[6] for an SPL in the HIS domain inspired by [20].

3 XML-Based Textual Feature Modeling

The idea of XML-based FM is not new and was originally proposed by Cechticky *et al* [21]. In that approach, the authors propose a XML-based backbone for a generative environment (please c.f. XFeature [22] derived from this concept). Though the purpose of that work is to use XML to define some models (SPL meta-meta-model, meta-model, and application meta-model, as well as other constraints models and display models) to allow user to define specific feature model and application model graphically, the underlying XML enabler techniques and concepts demonstrate the full power of XML in FM, which inspired us to uncover and reexamine them in the context of textual FM.

To make our XML-based approach intuitive, clear, and neutral, we try to use the basic XML elements to illustrate it only focusing on pure feature model elements without any display or tool specific tags.

3.1 Basic Models Elements

In this subsection, we show the XML modeling method for a feature model and leave the global constraints expression in the next subsection because the importance and complexity of constraints make it deserve a separate space.

Intuitively, we map a *feature* to an **element**, its cardinality to the occurrence time of the element, and its *attribute* feature(s) (if has) to **attribute**(s) of the element. In addition, *grouped features* are organized in a **sequence** (in which every element

[6] Compared to the original feature model proposed in [20], the adapted feature model not only uses the CMF notation, but also eliminates the confusion caused by combination of mandatory features and alternative features (e.g., the *ADSL* feature).

with a **minOccurs="0"** and let a validation checker (in our case, a Schematron[7] rule) to assure the cardinality of the group) or **choice** (can be used to model the feature group of children of *Internet connection*) for a shortcut of group cardinality of <1..1>. To save space, the code of Listing 1 shows the expression fragments for feature *Services* (Fig.1) in Schema with Schematron extension (model elements are in bold and syntax constructs are in boxes with background in gray. The group cardinality is complementally assured in line 10-11).

Listing 1

```
<xs:element name="Services">
  <xs:complexType>
   <xs:sequence>
     <xs:element ref="Video-on-demand" minOccurs="0"/>
     <xs:element ref="Internet-connection" minOccurs="0"/>
   </xs:sequence>
    <xs:attribute name="PRICE" type="xs:integer"/>
  </xs:complexType>
  <xs:annotation>
   <xs:appinfo>
    <sch:pattern>
     <sch:rule context="Services">
       <sch:assert test="fn:count(*)&gt;=1"/>
     </sch:rule>
     <sch:rule context="Services">
         <sch:assert test="@PRICE= fn:number(Video-on-demand@PRICE)
       +fn:number(Internet-connection@PRICE)"/>
     </sch:rule>
    </sch:pattern>
   </xs:appinfo>
  </xs:annotation>
</xs:element>
<xs:element name="Video-on-demand">
  <xs:complexType>
   <xs:attribute name="PRICE">
    <xs:simpleType>
     <xs:restriction base="xs:integer">
       <xs:minInclusive value="80"/>
       <xs:maxInclusive value="100"/>
     </xs:restriction>
    </xs:simpleType>
   </xs:attribute>
  </xs:complexType>
</xs:element>
```

Via this mapping, we can bi-transform a feature model automatically between its FD format and textual format to allow users choose their preference.

[7] http://xml.ascc.net/resource/schematron/

Please note, the new XML Schema 1.1 [38] supports both grammar checking (case for 1.0) and rule-based checking with <asset> element, as well as other enhanced semantics and syntax expressions. For instance, we can directly place `<assert test="XPath"/>` (the syntax is similar with that of in a Schematron rule) at the bottom of element's complex type declaration instead of introducing Schematron rules under the verbose `<annotation>` elements. But, considering the acceptance of 1.0, we still use the syntax of 1.0 with Schematron to illustrate the usage and indicate the difference when necessary.

3.2 Global Constraints Expression

Global constraints are essential to assure feature combination in a valid way. Though much literature (e.g., [4]) addresses its importance, few FM tools support complex constraints. In this aspect, a textual modeling method is more preferable and easier than a graphical one to represent the complex expressions.

In our example (Fig.1), feature *appliance control* must exist when *Power line* exists in a product. To express this constraint, we use another Schematron rule shown in Listing 2.

Though the constraint of our example is simple, it is only used here for an illustration purpose. In the XML community, it is well known that Schema plus Schematron can process any complex constraints for XML, like co-constraints, conditional constraints [23, 24, 25]. Other alternatives for advanced constraints are the combination of Schema and XSL [26, 27] or Schema 1.1. All approaches leverage XPath and are definitely also XML-based.

In Section 4, we can still see some other more details of XML for advanced features to support textual FM.

Listing 2

```
<xs:element name="Power-line">
 ...
 <xs:annotation>
  <xs:appinfo>
   <sch:pattern>
    <sch:rule context="/HIS/Control/Power-line">
     <sch:assert test="/HIS/Control/appliance-control"/>
    </sch:rule>
   </sch:pattern>
  </xs:appinfo>
 </xs:annotation>
</xs:element>
```

3.3 Staged Configuration

Staged configuration [16] is a process that allows the incremental configuration of (cardinality-based) feature models to be used in deferent stages of a development process or different organizations of a supply-chain. This is aligned with *stepwise refinement*, a fundamental approach to software development [28].

Suppose we want to limit the options of *Internet connection* to *ADSL* and *Wireless* in some areas, we can implement this configuration using the standard <redefine> (<override> in Schema 1.1) element. The code of Listing 3 demonstrates our needs.

Listing 3

```
<xs:redefine schemaLocation="HIS.xsd">
  <xs:complexType name="Internet-connection-Type">
    <xs:choice>
      <xs:element ref="ADSL"/>
      <xs:element ref="Wireless"/>
    </xs:choice>
  </xs:complexType>
</xs:redefine>
```

An alternative approach is to use an extended tool (maybe) based on XAK [29] to allow elements removal and fine-grained control on attributes for XML refinements.

4 Evaluation and Discussion

In the preceding section, we show how a general feature model to be expressed in pure XML syntax in a feasible way. In this section, we will examine the XML-based textual FM approach in a more detailed way.

4.1 Evaluation Criteria

In [8], Classen *et al.* listed 8 criteria for comparing textual FM languages, they are: (i) *human readability*, i.e., whether the language is meant to be read and written by a human; (ii) support for attributes; (iii) decomposition (group) cardinalities; (iv) basic constraints, i.e., *requires*, *excludes* and other Boolean constraints on the presence of features; (v) complex constraints, i.e., Boolean constraints involving values of attributes; (vi) mechanisms for structuring and organizing the information contained in a FM (other than the FM hierarchy); (vii) formal and tool-independent semantics and (viii) tool support.

Within our knowledge, this criteria list is the only published in this field (though general languages evaluation criteria exist, e.g., [35]). To make the XML-based method comparable with the languages listed in [8] (especially TVL), we evaluate the XML-based method in these aspects (other criteria will be discussed later in this section).

4.2 Evaluation Results

The evaluation of the XML-based approach is performed in three small case studies with three development teams of an industry company. Each team has 4 to 6 technical staff (architect or developer) who have fairly knowledge on XML but not necessary for SPL. Each team was interviewed individually with one or two of the authors as tutor(s). Each case study lasts three days (a half day for XML-based textual FM introduction, 2 days for feature model construction with the help of tutor(s) (if possible), a half day for final evaluation as interview between the team and the tutor(s)). The feedbacks were collected according to the overall discussion of each team on the 8 aspects listed in the previous subsection. TABLE I shows the evaluation results, as well as the mechanisms of XML on the evaluated aspects. The evaluation scales in Table 1 are: ■ strongly satisfied; + rather satisfied; ○ neither satisfied nor unsatisfied; — rather unsatisfied.

These results show that the XML-based approach is compatible with TVL in all these aspects, since all these aspects are also supported with different level in TVL [8], though *human readability* and tool support are not perfect in our approach (we will discuss them later in this section).

Table 1 Evaluation Results of the XML-based Method

Cri.	XML-based FM	
	Mechanisms	**Level**
i	XML tags, depends on the famality of user to XML	+, ○, ○
ii	attribute element; also can be modeled as element with simple or complex type	■, +, +
iii	Group tags, like `<sequcence>`, `<choice>`, with Schematron rules (or `<assert>` in Schema 1.1)	■, +, +
iv	Schematron rules; XSL/XSLT	■, ■, +
v	Schematron rules; XSL/XSLT	■, +, ○
vi	`<include>` and `<import>` directives	■, ■, +
vii	Formal semantics of XML (Schema)	■, +, +
viii	XSD or XML tools (e.g., XML Spy)	+, —, —

In addition, we get more encouragements with regards to considering deeper and finer grained aspects which are important for industry adoption. For instance, in the language level of features and constructs, XML supports:

- **Attribute cardinality**. [0..1] for attributes (in XML) and [*n*..*m*] for elements (if the attribute feature is modeled as an element in XML).
- **Cloning**. Via feature cardinality.
- **Default values**. Only for attribute feature.
- **Extended type set**. Besides the 19 (more with 1.1) built-in primitive types provided by Schema, users can derive (restrict or extend) them, use them in lists or unions, and even define new ones thanks to the extensibility of Schema.
- **Import mechanism**. `<import>` for more fine-grained scoping control beyond `<include>`.
- **Specialization**. `<redefine>` (`<override>` in 1.1) or refinement techniques like XAK for staged configuration.

But TVL does not support all these language level features except that only four data types (`enum`, `int`, `real`, `bool`) are supported in current TVL [30, 8].

4.3 *Findings and Observations*

Compared to TVL, the main advantages of XML are:

1. **Completeness of language.** XML and related standards are formally defined with semantics which totally amount to hundreds pages in print. This richness in languages make XML have full functionality in FM, which is also confirmed by the smoothness of FM in all the three case studies.
2. **Openness and Extensibility.** XML is designed for openness and extensibility which allow us to design very flexible Schemas for the purposes of reuse, variation, substitution, etc. And we can even design a totally new textual FM language based on XML, i.e., a domain specific language (DSL) for FM.
3. **Standard-based portability.** XML is widely used in most of (if not all) data processing domain as a neutral data format. Therefore, it is well recognized in most tools and applications, which means that it has a high portability, though specific semantics for them are to be specified.
4. **Simplicity and easy to learn.** Though XML and Schema have complex semantics, the use of them for FM is fairly simple with very few model elements mapping, which makes this approach easy to learn for users familiar with XML and (few) FM concepts.
5. **Vast acceptance and huge user base.** As a standard, XML has been being used for more than a dozen of years and has penetrated into nearly all areas of IT. This universality makes XML has a huge user base and is popular among business and technical staff.

However, there are also some limitations of XML:

1. **Lower clarity and expressiveness.** As a general data description language, XML has many (maybe) irrelevant tags for FM, which makes the text verbose and hard to read. This is evident from the example codes in Section II (the maybe irrelevant sentences, compared to a DSL, occupy 1/3 space (see the sentences in gray).). Also, the use of *element*s to indicate *features* is not of clarity of purpose, though such direct mapping from *feature*s to *elements* is natural.
2. **Lower abstraction** (especially in constraints expression). This aspect has some overlapping with the previous one for model elements definition. The main abstraction problem pointed out by two of the three interviewed teams (and also pointed in [24]) is the complex constraints expression introduced by Schematron and they had to learn this rule language. With regard to Schema 1.1, the user still needs to learn XPath functions and expressions to write a complex constraint. Therefore, a language with higher level of abstraction is preferable.
3. **Naive tool support.** Though the text-based nature allows XML to be written in any text editors, the XML feature model can only be treated as an ordinary XML with validation assurance posed by Schema. There are no tools to support such XML feature model analysis, e.g., to ensure a specialized feature model still obey the feature constraints posed by the original one in a staged configuration. This is not surprising since there is none explicit textual XML FM approach, not to mention tools.

These issues and limitations of XML lead to the following discussions to improve the XML-based textual FM method with some possible techniques.

4.4 Discussion

4.4.1 Readability and Clarity of XML

A main limitation referred to XML is the verbose format, as well as some tags (e.g., the `<complexType>` tag) seemed irrelevant but have equal position in visualization as those key ones. This is true since XML does not have any knowledge on feature models, and this problem is not specific to XML, but any languages with declarative markups, e.g., C preprocessors. This problem was addressed recently by some emerging visual techniques in SPL community, e.g., color annotation [33] and zoomable viewer [32], which use different visual effects to differ different elements to make the interested ones are more visible to the users via a virtual view of separation of concerns [31]. All this techniques can be used to improve the clarity of XML. Another issue related to readability is the domain (FM) concepts sensitivity for the key words used in XML, which is involved in the next issue.

4.4.2 XML-Based DSL Vs. General Purpose Language (GPL)

"Is not it better to replace `<element>` with (<) **`feature`** (>)?" Yes, this is definitely true for making domain semantics explicit and it is the main advantage of a DSL over a GPL [34]. Also, the solution to this issue can benefit the previous one. But we do not think it is a good candidate to have a totally new DSL with XSD customization since it is a very hard and tedious work needing good knowledge on XSD (validation) implementation. A better option is to use a simplified and domain-oriented view (physical or virtual) to hide the underlying XML representation, which can be achieved via a model to model transformation. Another improvement of DSL is to use high level interfaces with proper abstraction to hide the complexity of Schematron rules for constraints. This can be done via customized XPath functions, which provides seamlessly integration with underlying programming environment to leverage the functionality of a programming language. This capability is important for modeling [36] and also an advantage of XML.

4.4.3 XML-Based Feature Analysis and Reasoning

Feature analysis and reasoning are widely used as a diagnosis technology in the process of product configuration or testing. There are still challenges in this field for large models [7], but this has noting to do with the representation of XML for that we can translate any (in theory) feature models into symbols of CSP, SAT, or BDD though their combination for efficient reasoning is still a open question [7]. However, the rule-based nature of Schematron and XSL provide yet another implementation for reasoning [37] though similar with CSP in concept but deserve deeper analysis and comparison.

4.4.4 Tool Support

All the issues above are related to, in some level, the support of tools. This is a key factor to its eventual adoption in industry. Besides the techniques discussed before, the basic requirements for syntax highlight, code assistant and completion must be supported in a modern programming environment. Both XML editor and TVL are in a very early stage with regards to this aspect. However, we can leverage many existing open source components (e.g., Sat4j[8], Schematron plugin[9], model transformation toolkit[10], DSL framework[11], rule engine[12], etc.) into a harmonious integrated environment like Eclipse.

4.4.5 Points of View

Effective representations of feature models are an open research question [30]. The XML-based textual FM approach inherits many outstanding characteristics of XML like rich data types, attributes declaration, modularization, and openness which are essential for feature models scalability in real industry settings. The full support for any complex constraints expression and validation and seamlessly integration with programming environment also make it a better (if not best) candidate for FM. Above all, the vast of acceptance in nearly every areas of IT and popularity among both technical and business staff also play a very important role in its adoption in the industry and should never be underestimated.

In the emerging textual FM field, TVL provides a DSL which is more clear and intuitive than pure XML. However, the current ability of TVL is not superior to XML with regards to textual FM in language level. Though the authors of TVL stated that more functionality will be added to TVL, how to protect it from becoming more complex and not pure as it is? If so, why not create a DSL on a mature platform with less effort?

As researchers, we are interested in new techniques creation. But as practitioners, the more important things for use are to choose the right technologies and tools and combine them in an economical and pragmatic way which involves technology maturity, learning cost, technical support and long term development which are often neglected by the academic.

We acknowledge that the tool support for XML-based textual FM is still naive. But we cannot blame XML because whether a technology is good or bad depends on how to use it. As a general data processing and modeling platform, XML has already proved it in a FM and generative environment [21]. Just as the graphical notations use UML as a modeling platform, the counterpart in text notations is XML because a natural mapping can be built between UML (MOF) and XML (Schema). So, we have confidence to believe that the XML platform can play a more important role in textual FM if we improve them in a right way.

---

[8] http://sat4j.org/
[9] http://www.ibm.com/developerworks/cn/opensource/os-cn-eclipse-xmlvalid1/index.html
[10] http://eclipse.org/atl/
[11] http://www.eclipse.org/Xtext/
[12] http://www.jboss.org/drools

5 Threats to Validity

Let us briefly discuss the threats to validity of our work.

Rigorousness of the XML-based FM approach. The main purpose of the XML-based textual FM approach introduced in this paper is to give an intuitive illustration to demonstrate the natural support of XML in this field. Thus we do not confine a specific implementation since users can define more advanced models via meta-modeling.

Feature modeling adoption in the industry. FM is still rarely adopted in the industry though widely used in the academic. We do not think the XML-based approach can play a decisive role since many factors involved, e.g., domain maturity, but a better tool helps.

Empirical studies for XML and TVL comparison. We have not conducted some empirical studies to compare the XML-based approach and TVL, for one reason that TVL is a rather new published language, for the other that both approaches are in a very early stage with naive tool support. The main purpose of "comparison" between them in language level is to show XML has compatible abilities with TVL to recall the reexamination of XML among practitioners. However, an empirical study would be necessary when both have better tool support.

6 Conclusions and Future Work

Feature modeling is an essential activity for SPLE and has a strong influence in SPL adoption in the industry. A textual FM method or language can make up the limitations of the graphical ones and combines it to facilitate FM practices. To make the textual FM approach pragmatic, we propose the XML-based approach, demonstrate its power with a partial HIS SPL, and reveal its advantages with comparison to another textual FM language, TVL. Though the XML-based approach still has shortcomings like readability and more advanced tool support, we believe all this problems can be solved using the ideas and techniques discussed in this paper. Due to the vast acceptance of XML in the industry, we hope the XML-based approach can play a more important role in languages design for FM.

We are currently working towards a smarter tool for improving the readability and enforcing separation of concerns for textual feature models based on XML. Also, we are also planning to conduct some empirical studies for the XML-based textual FM and compare it with TVL in a more rigorous way.

Acknowledgments. The authors thank the anonymous referees for their helpful comments on an earlier version of this paper. This work was partially supported by the National Basic Research Program of China under Grant Nos. 2012CB724107, 2010CB735907.

References

1. Clements, P., Northrop, L.: Software Product Lines: Practices and Patterns. Addison-Wesley (2002)
2. Pohl, K., Böckle, G., van der Linden, F.: Software Product Line Engineering. Springer, Heidelberg (2005)
3. Czarnecki, K., Eisenecker, U.: Generative Programming: Methods, Tools, and Applications. Addison-Wesley (2000)
4. Kang, K.C., Cohen, S.G., Hess, J.A., Novak, W.E., Peterson, A.S.: Feature-Oriented Domain Analysis (FODA) Feasibility Study. Technical Report CMU/SEI-90-TR-21, ESD-90-TR-222 (1990)
5. El Dammagh, M., De Troyer, O.: Feature Modeling Tools: Evaluation and Lessons Learned. In: De Troyer, O., Bauzer Medeiros, C., Billen, R., Hallot, P., Simitsis, A., Van Mingroot, H. (eds.) ER Workshops 2011. LNCS, vol. 6999, pp. 120–129. Springer, Heidelberg (2011)
6. Antkiewicz, M., Czarnecki, K.: Feature Plugin: Feature Modeling Plug-In for Eclipse. In: Proc. 2004 OOPSLA Workshop on Eclipse Technology Exchange, pp. 67–72. ACM Press (2004)
7. Batory, D., Benavides, D., Ruiz-Cortés, A.: Automated Analyses of Feature Models: Challenges Ahead. Communications of the ACM 49(12), 45–47 (2006)
8. Classen, A., Boucher, Q., Heymans, P.: A text-based approach to feature modelling: Syntax and semantics of TVL. Science of Computer Programming 76(12), 1130–1143 (2011)
9. van Deursen, A., Klint, P.: Domain-Specific Language Design Requires Feature Descriptions. Journal of Computing and Information Technology 10(1), 1–17 (2002)
10. Batory, D.: Feature Models, Grammars, and Propositional Formulas. In: Obbink, H., Pohl, K. (eds.) SPLC 2005. LNCS, vol. 3714, pp. 7–20. Springer, Heidelberg (2005)
11. Boucher, Q., Classen, A., Faber, P., Heymans, P.: Introducing TVL, a Text-based Feature Modelling Language. In: Proc. Int. Workshop on Variability Modelling of Software-Intensive Systems (VaMoS 2010). ICB-Research Report 37 Universität Duisburg-Essen, pp. 159–162 (2010)
12. Czarnecki, K.: Variability Modeling: State of the Art and Future Directions. In: Proc. Int. Workshop on Variability Modelling of Software-Intensive Systems (VaMoS 2010), ICB-Research Report 37 Universität Duisburg-Essen, p. 11 (2010)
13. Benavides, D., Segura, S., Trinidad, P., Cortés, A.R.: FAMA: Tooling a Framework for the Automated Analysis of Feature Models. In: Proc. Int. Workshop on Variability Modelling of Software-Intensive Systems (VaMoS 2007), Lero Technical Report 2007-01, pp. 129–134 (2007)
14. Kang, K.C., Kim, S., Lee, J., Kim, K.: FORM: A Feature-Oriented Reuse Method. Annals of Software Engineering 5(1), 143–168 (1998)
15. Griss, M., Favaro, J.: dlAlessandro, M.: Integrating Feature Modeling with the RSEB. In: Proc. Fifth Intl. Conf. on Software Reuse, ICSR 1998, pp. 76–85. IEEE Press (1998)
16. Czarnecki, K., Helsen, S., Eisenecker, U.W.: Formalizing cardinality-based feature models and their specialization. Software Process: Improvement and Practice 10(1), 7–29 (2005)

17. van Gurp, J., Bosch, J., Svahnberg, M.: On the Notion of Variability in Software Product Lines. In: Proc. Working IEEE/IFIP Conference on Software Architecture, WICSA 2001, pp. 45–54. IEEE Press (2001)
18. Riebisch, M., Bollert, K., Streitferdt, D., Philippow, I.: Extending Feature Diagrams with UML Multiplicities. In: Proc. IDPT 2002 (2002)
19. Benavides, D., Trujillo, S., Trinidad, P.: On the Modularization of Feature Models. In: Proc. First European Workshop on Model Transformation (2005)
20. Benavides, D., Trinidad, P., Ruiz-Cortés, A.: Automated Reasoning on Feature Models. In: Pastor, Ó., Falcão e Cunha, J. (eds.) CAiSE 2005. LNCS, vol. 3520, pp. 491–503. Springer, Heidelberg (2005)
21. Cechticky, V., Pasetti, A., Rohlik, O., Schaufelberger, W.: XML-Based Feature Modelling. In: Dannenberg, R.B., Krueger, C. (eds.) ICOIN 2004 and ICSR 2004. LNCS, vol. 3107, pp. 101–114. Springer, Heidelberg (2004)
22. Pasetti, A., Rohlik, O.: Technical Note on a Concept for the XFeature Tool (June 15, 2005), `http://www.pnp-software.com/XFeature/pdf/XFeatureToolConcept.pdf`
23. Costello, R. L.: Best Practice: Use Multiple Schema Languages (November 1, 2006), `http://www.xfront.com/Integrated-schema-approach/Use-Multiple-Schema-Languages.html`
24. Costello, R.L.: Extending XML Schemas (2006), `http://www.xfront.com/BestPracticesHomepage.html`
25. Marinelli, P.C., Coen, S., Vitali, F.: SchemaPath, a Minimal Extension to XML Schema for Conditional Constraints. In: Proc. Int. Conference on World Wide Web, WWW 2004, pp. 164–174. ACM Press (2004)
26. Domínguez, E., Lloret, J., Rubio, Á.L., Zapata, M.A.: Validation of XML Documents: From UML Models to XML Schemas and XSLT Stylesheets. In: Yakhno, T., Neuhold, E.J. (eds.) ADVIS 2006. LNCS, vol. 4243, pp. 48–59. Springer, Heidelberg (2006)
27. Heneback, P.: Advanced XML validation (2006), `http://www.ibm.com/developerworks/xml/library/x-crsfldvalid/`
28. Wirth, N.: Program Development by Stepwise Refinement. Communications of the ACM 14(4), 221–227 (1971)
29. Anfurrutia, F.I., Díaz, Ó., Trujillo, S.: On Refining XML Artifacts. In: Baresi, L., Fraternali, P., Houben, G.-J. (eds.) ICWE 2007. LNCS, vol. 4607, pp. 473–478. Springer, Heidelberg (2007)
30. Hubaux, A., Boucher, Q., Hartmann, H., Michel, R., Heymans, P.: Evaluating a Textual Feature Modelling Language: Four Industrial Case Studies. In: Malloy, B., Staab, S., van den Brand, M. (eds.) SLE 2010. LNCS, vol. 6563, pp. 337–356. Springer, Heidelberg (2011)
31. Kästner, C., Apel, S.: Virtual Separation of Concerns – A Second Chance for Preprocessors. Journal of Object Technology 8(6), 59–68 (2009)
32. Stengel, M., Feigenspan, J., Frisch, M., Kästner, C., Apel, S., Dachselt, R.: View Infinity: A Zoomable Interface for Feature-Oriented Software Development. In: Proc. Int. Conf. Software Engineering, ICSE 2011, pp. 1031–1033. ACM Press (2011)

33. Feigenspan, J., Schulze, M., Papendieck, M., Kästner, C., Dachselt, R., Köppen, V., Frisch, M.: Using Background Colors to Support Program Comprehension in Software Product Lines. In: Proc. International Conference on Evaluation and Assessment in Software Engineering, EASE 2011, pp. 66–75. IET Press (2011)
34. Sprinkle, J., Mernik, M., Tolvanen, J., Spinellis, D.: What Kinds of Nails Need a Domain-Specific Hammer? IEEE Software 26(4), 15–18 (2009)
35. Holtz, N.M., Rasdorf, W.J.: An evaluation of programming languages and language features for engineering software development. Engineering with Computers 3(4), 183–199 (1988)
36. Völter, M.: From Programming to Modeling—and Back Again. IEEE Software 28(6), 20–25 (2011)
37. Costello, R.L.: Generating New Knowledge by Deductive Reasoning using Schematron (2010), http://www.xfront.com/Generating-New-Knowledge-by-Deductive-Reasoning-using-Schematron.pdf
38. Gao, S., Sperberg-McQueen, C.M., Thompson, H.S.: W3C XML Schema Definition Language (XSD) 1.1 Part 1: Structures. W3C Proposed Recommendation (2012), http://www.w3.org/TR/xmlschema11-1/

Dealing with NFRs for Smart-Phone Applications: A Goal-Oriented Approach

Rutvij Mehta, Hongyuan Wang, and Lawrence Chung

Abstract. The utility of a smart-phone application depends not only on its functionality but also on its key non-functional requirements (NFRs), such as ubiquity, safety and usability. Omissions or commissions of considerations of such NFRs could lead to undesirable consequences, such as lack of user safety and difficulty in using smart-phone features. Currently, however, there is little systematic methodology for dealing with NFRs for a smart-phone application, in consideration of the particular characteristics of smart-phones, such as limited screen-size and battery-life, and the availability of a variety of embedded sensors and input/output devices. In this paper, we propose a goal-oriented approach in which NFRs are treated as softgoals, and then used in exploring, and selecting among, alternative means for satisficing them. In this approach, both synergistic and antagonistic interactions among the softgoals are identified and analyzed, considering the particular characteristics of smart-phones. As a proof of concept, a fall detection and response feature of a smart-phone application is presented, along with a safety scenario.

Keywords: Goal-oriented, Non-Functional Requirements (NFRs), Softgoal Interdependency Graph (SIG), Smart-phone applications, Ubiquity, Safety, HOPE (Helping Our People Easily).

1 Introduction

Use of smart-phone applications is increasingly becoming an indispensible part of everyday life. Smart phones have certain specialized characteristics in terms of

Rutvij Mehta · Lawrence Chung
Department of Computer Science, The University of Texas at Dallas, Richardson, TX, 75080, USA
email: rutvij.mehta@utdallas.edu, chung@utdallas.edu

Hongyuan Wang
Department of Computer Science, Jilin University, Changchun, Jilin, 130012, China
email: hongyuan@jlu.edu.cn

R. Lee (Ed.): Software Engineering Research, Management and Appl. 2012, SCI 430, pp. 113–125.
springerlink.com

screen size, battery power, embedded sensors and such. These characteristics differentiate smart-phones from traditional devices concerning functionality and usage. In order to be effective, however, a smart-phone application should be considered not only in terms of its functional but also non-functional requirements (NFRs), such as ubiquity, safety, power conservation, among others. It therefore seems noteworthy to consider various characteristics such as battery power, screen size, in evaluating NFRs for smart-phone applications. Fig. 1 depicts some of the characteristics.

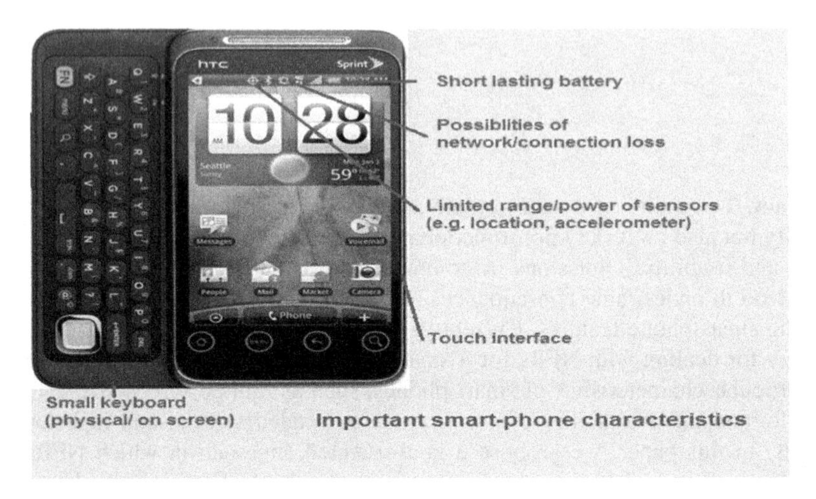

Fig. 1 Important smart-phone characteristics.

Currently, there is little systematic methodology, for considering key NFRs such as ubiquity and safety for smart-phone applications. In particular, omissions of such considerations could lead to undesirable consequences, such as lack of user safety and exposure of personal information, apart from an unusable product, but have not been addressed adequately in the past.

In this paper, we propose to treat NFRs for smart-phone applications as softgoals to be satisficed. This treatment then enables us to take a goal-oriented approach, in which the notions of NFRs can be successively refined and operationalized. In the goal-oriented approach, alternatives can be explored and selected, while the various synergistic and antagonistic interactions among them are identified and analyzed, considering the properties and nature of a smart-phone environment.

Using an emergency fall detection and response scenario, we show both intra-NFR analysis for the interactions among various alternatives for a particular NFR and inter-NFR-and-other-NFRs analysis for the interaction among more than one NFR, in particular, using ubiquity, safety, and long battery life.

The rest of the paper is organized as follows. An emergency scenario involving elderly fall detection and response is discussed in section 2. Section 3 describes how to define and analyze NFRs as a softgoals (e.g. ubiquity as a softgoal), and consequently use a goal-oriented rational approach to systematic decision making. It also discusses the *intra-NFR* and *inter-NFR-and-other NFRs* analyses. Section 4 describes our prototype smart-phone system, the HOPE system, together with our goal-oriented implementation. In light of the initial observations, Section 5 describes some related work, while Section 6 gives a summary of conclusions and future work.

2 An Emergency Scenario

HOPE (Helping Our People Easily) is a smart-phone application we have been developing in the past several years, where safety is one of the key non-functional softgoals, along with ubiquity, usability and others such as power-saving and cost. This scenario involves an elderly person who is subject to a fall (could be at home or some other place), and due to the age-related symptoms, such as speech impairment or lack of dexterity, is unable to communicate her situation to others. This may also prove fatal. The problem view of this scenario is seen in Fig. 2.

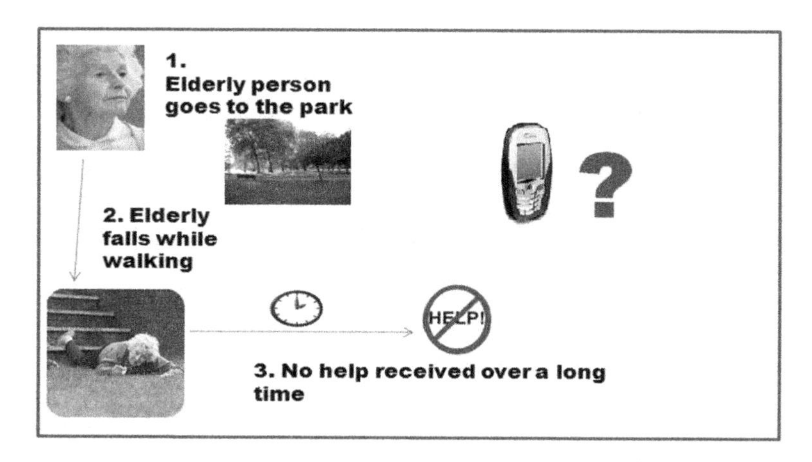

Fig. 2 An emergency scenario (A problem view).

Some of the possible actions include, simply waiting for someone to notice and help – that may not be timely, a body embedded sensor that may prove costly and invasive, a smart-phone application that senses the fall and sends out location and alert, among others. Assuming we select using a smart-phone to detect a fall and send out an alert, there still arise questions concerning NFRs that may prove challenging. Few of the questions could be:

a) How to best detect this situation and help the user? (**Ubiquity, Safety**)
b) Should GPS be kept on at all times? (**Longer battery life**)
c) How to make it easy for user to choose the individualized action? (**Usability**)

The monitoring and detection of this situation in the most effective manner, along with the best possible action to take, involves the study of goal-orientation. Ubiquity and Safety become important non-functional softgoals as indicated.

3 Our Goal-Oriented Approach

As seen from the previous scenario description, NFRs play an important role in modeling a solution. However, lack of a systematic process for analyzing such questions for dealing with NFRs could lead to an ineffective solution. Some of the research challenges lie in answering such questions:

- How to address some key NFRs such as Ubiquity, and explore desirable properties for these attributes, for smart-phone applications?
- How to deal with interactions among NFRs? e.g. How ubiquity, affects another closely related attribute, say safety

Our goal-oriented approach to deal with NFRs for smart-phone applications extends the NFR framework approach deal with NFRs. We now introduce steps for applying our goal-oriented framework for smart-phone applications:

- Post the NFRs such as ubiquity, safety as softgoals.
- Outline the alternative actions for each soft goal, considering the characteristics of smart-phones.
- Analyze these alternatives, with respect to their tradeoffs, i.e., positive or negative impacts on the NFR softgoals.
- Perform both the intra-NFR as well as inter-NFR-and-other-NFRs analysis according to the above steps.
- Prioritize the softgoals throughout the process of disambiguation, operationalization and conflict resolution.
- Select the best possible alternative(s).

In order for the above, we extend the NFR Framework [1], which is intended to help address NFRs systematically and rationally. The NFR Framework facilitates the exploration and analysis of, as well as selection among, alternatives, through the use of a visual notation, called SIG (Softgoal Interdependency Graph [1]. A SIG consists of such ontological concepts as non-functional softgoals (e.g., Ubiquity), operationalizing softgoals (e.g., Detect location) and contributions (e.g., Make, Break, Help and Hurt) that one softgoal has towards another. In concept, SIG is built over the AND-OR graph, consisting of the 'And' (all sub-goals need to be satisficed) as well as 'Or' (one sub-goal satisficing leads to the satisficing of its parent goal) decomposition. The label propagation allows us to visually depict whether a softgoal is satisficed/denied by their alternative actions. Similarly, the soft-goals/operationalizations can be prioritized for criticality. These mechanisms are essential in conflict resolution. In this paper, we construct the SIGs using [16]. First, we outline the subjectivity in defining and refining NFRs in terms of alternatives, and then present the interactivity property (both the intra-NFR as well as inter-NFR-other-NFRs analyses).

3.1 Subjectivity of NFRs for Smart-Phone Applications

Subjectivity is an important attribute for NFRs. This subjectivity is seen across levels such as in defining a softgoal, refining the softgoal in terms of alternatives as well as in terms of operationalizations or mechanisms to implement the alternatives. We shall see the subjectivity applied to the softgoal of *ubiquity* in consideration of smart-phone applications. The idea of ubiquitous computing has been synonymous with 'invisible computing' [2]. Over the years, researchers have identified various topics such as context-awareness, availability, mobility, adaptability and so on, as being associated with ubiquitous computing. Many researchers have outlined important attributes of ubiquity as **<Availability, Invisibility, Tolerate Ignorance, Mobility, Context-Awareness>**, among others. However, many of these works differ in ways to define ubiquity. Few works, who have outlined their attribute lists, have not treated 'ubiquity' as a softgoal. Two of these works [3, 4], have been modeled here using our approach, as shown in Fig. 3 and 4 respectively.

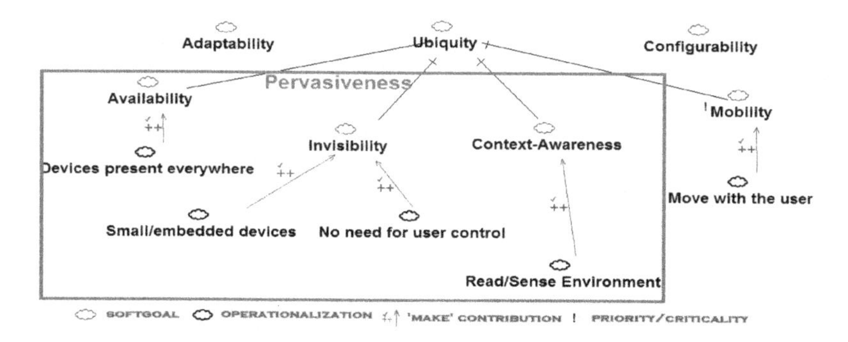

Fig. 3 Treating ubiquity as a softgoal, using our approach.

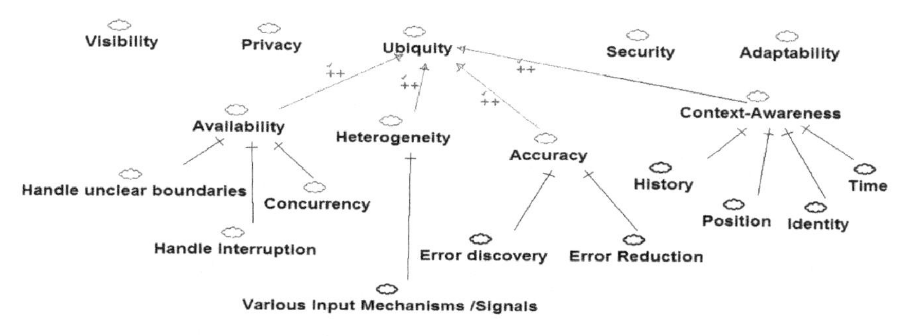

Fig. 4 Alternative refinement of the ubiquity softgoal.

To further clarify, our approach enables us to take various alternative ways of defining and refining the same softgoal to deal with the subjectivity of NFRs. Again, the previous works [3, 4], do not follow a goal-oriented approach and have just presented their own list of attributes. The significance of our approach is demonstrated in the fact that subjectivity can be handled, irrespective of the application. This approach also allows us to define relationships among alternatives and softgoals which are clarified in section 2.2.

3.2 Interactivity of NFRs for Smart-Phone Applications

Interactivity is another important attribute of NFRs. Interactivity is characterized in terms of both synergistic and antagonistic relationships. This property is demonstrated in our approach, in two levels such as intra-NFR (considering one softgoal) as well as inter-NFR-and-other-NFRs (more than one softgoal analyzed together). We shall demonstrate both these levels modeling ubiquity, safety and long battery life softgoals using our emergency scenario.

3.2.1 Intra-NFR Analysis

As seen from Fig. 3 and 4, and the goal-oriented approach, there can be various ways in which ubiquity can be defined and refined. We present *our* definition of ubiquity concerning the emergency scenario in Fig.5 using the concepts from both these definitions along with extending the concepts illustrated in [3, 4].

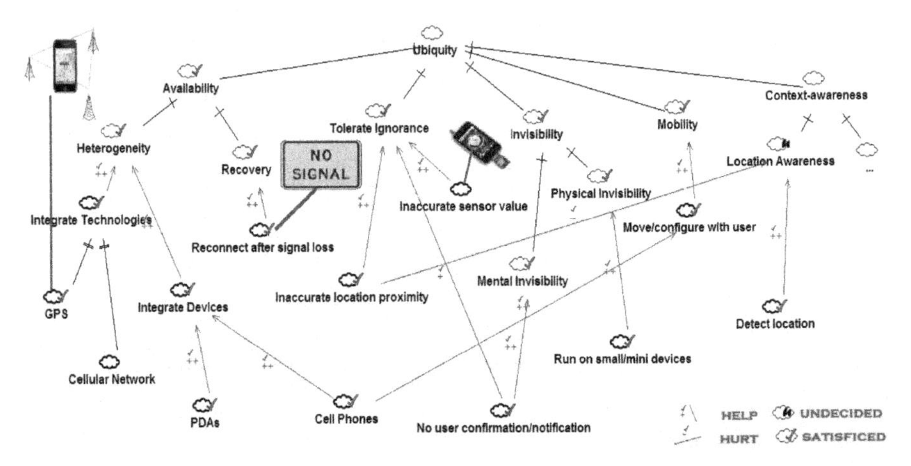

Fig. 5 A SIG showing intra-ubiquity analysis for the emergency scenario.

Note that this is just one possible way of defining and refining ubiquity. There can be another set of refinements and alternatives for the ubiquity softgoal in the context of a particular application. Some alternatives for this scenario, along with the 'topic' component of the SIG, are not shown in the above figure for clarity. The selection of alternatives is also influenced by other non-functional attributes (not shown here) such as cost and usability. One interesting sub-goal, 'Tolerate Ignorance' [5], is where approximations are taken into consideration rather than only relying for exact values. This is important as even though the sensor values (accelerometer that detects the fall), may be slightly inaccurate, the system must continue to function in the best possible manner. This is useful sometimes such as in the case of there not being an actual fall (false positive), the system sends out an alert. There might be a class of users, who would rather allow this behavior than for the system to miss out detecting a fall in waiting for accurate results.

3.2.2 Inter-NFR-and-Other-NFRs Analysis

We shall now deal with the interaction of ubiquity with two such non-functional requirements, safety and conserving (or longer) battery life, for our emergency scenario.

3.2.2.1 Inter-Ubiquity-and-Safety Analysis

As seen in the emergency scenario, safety is one of the key softgoals to be achieved. Safety, [6], in itself is an important non-functional goal that has been previously modeled by researchers in different ways. The interaction of safety along with ubiquity is demonstrated in the SIG presented in Fig. 6.

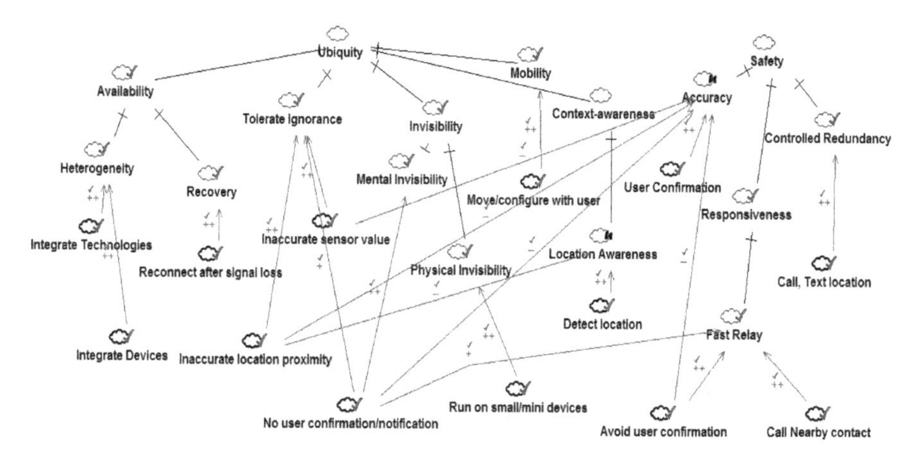

Fig. 6 A SIG showing inter-ubiquity-and-safety softgoal analysis.

We limit safety softgoal decomposition in accordance to the other top-level softgoal of ubiquity. For this situation, safety would be detecting a fall and placing the emergency call along with sending the user location, in order to get fast and effective response. It is important to note that for safety we shall be adopting the open world assumption, meaning the alternatives here are not the only ones that satisfy the softgoal. For our scenario, safety has sub-goals namely, **<Accuracy, Responsiveness, and Controlled Redundancy>**. Again, safety can be modeled for subjectivity as was seen earlier using ubiquity and therefore various definitions exist for defining this top-level softgoal.

3.2.2.2 Inter-Ubiquity-and-Long Battery Life Analysis

As illustrated earlier in Fig. 1, a relatively short battery life is one of the key characteristics of smart-phones. Such characteristics need to be studied in order to make a systematic decision in terms of selecting the operationalizations. In order to have a longer battery life, the user is faced with certain alternatives such as– 'Turn of power consuming sensors such as GPS', or 'Reduce the volume or screen brightness', or even 'Turn off background processes'. Revisiting the emergency scenario, which of these alternatives should the user select? A rational decision cannot be taken by considering the softgoal of 'Longer battery life' by itself. We need to study the effect (positive and/or negative) an alternative has on other inter-related softgoals. The 'ubiquity-long battery life' analysis is depicted in Fig. 7.

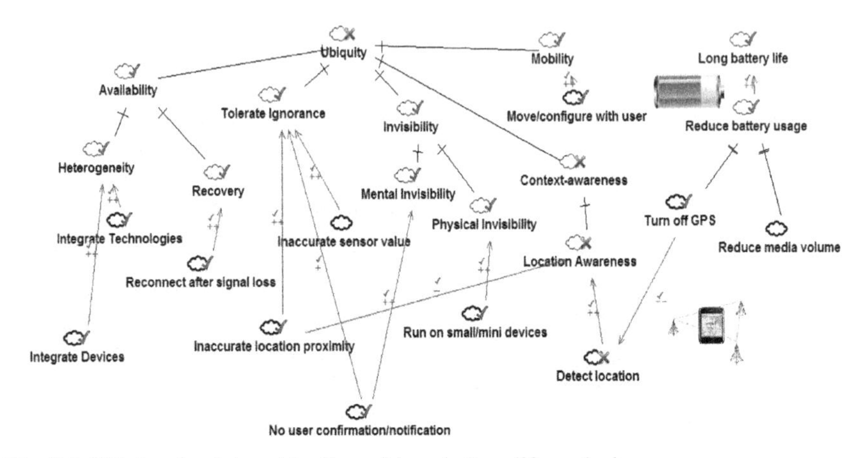

Fig. 7 A SIG showing inter-ubiquity-and-long battery life analysis.

We depict the interactivity using our approach by studying the softgoal of 'ubiquity' along with the softgoal of a 'long battery life'. We can observe that using GPS to detect location can cause a conflict with an alternative of preserving smart-phone battery. Again, interactivity should be studied with as many interacting softgoals as possible depending upon the nature of the application e.g. ubiquity-safety-privacy-long battery life.

4 Implementation

The HOPE software system [7] is designed to alleviate the communication issues faced by the elderly due to their age related symptoms such as reduced hearing, memory, and vision and speech loss. It is built on the Android OS mobile platform. It works on the principle that human perception is visual and includes icons and pictures. It also has text-to-speech and speech recognition capacities to help overcome the communication barriers. '*Fall detection and response*', which is discussed throughout the paper, is one sub-feature in the 'Emergency' feature present in the HOPE software system. The home screen of the HOPE system, showing various sub-features, is depicted in Fig. 8 below.

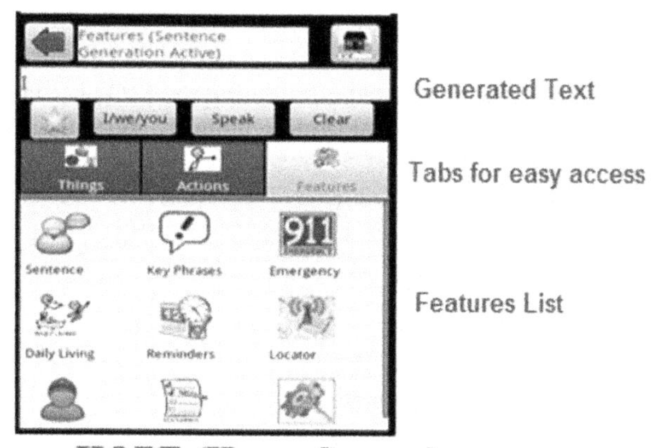

Fig. 8 HOPE System – Home Screen.

4.1 Trade-Off Analysis

Using our goal-oriented approach for dealing with NFRs for smart-phones, we can make a rational selection decision among these alternatives. However, there can still be conflicts among these alternatives (e.g. as seen in Fig. 7). We, therefore, use the prioritization mechanism for selection. Let us consider this situation – Suppose the user has the goal of conserving battery power and therefore turns of the GPS sensor.

(Goal – Conserve battery) -> (Action – Turn off GPS) (Conflict – Location not detected; affects Ubiquity and Safety)

In order to prevent this, we assign a higher priority to the softgoal of Safety.

(Prioritize – Safety!!)

(Action – Turn on GPS to detect location)
(Consequence – We can send location details to emergency contact to have a faster response)

Going further, we can even help conserve the battery, by turning on the GPS only in case of fall detection and turning it off after sending the alert.

(Prioritize – Safety!! And Conserve Battery!)

(Action – Turn on GPS to detect location only at the time fall is detected, turn it off after sending out the alert)
(Consequence – We can send location details to emergency contact to have a faster response and help conserve the battery)

4.2 Goal-Oriented Implementation

The emergency scenario is implemented considering the SIG alternatives, into the HOPE system as shown in Fig. 9, demonstrating ubiquity and safety softgoals.

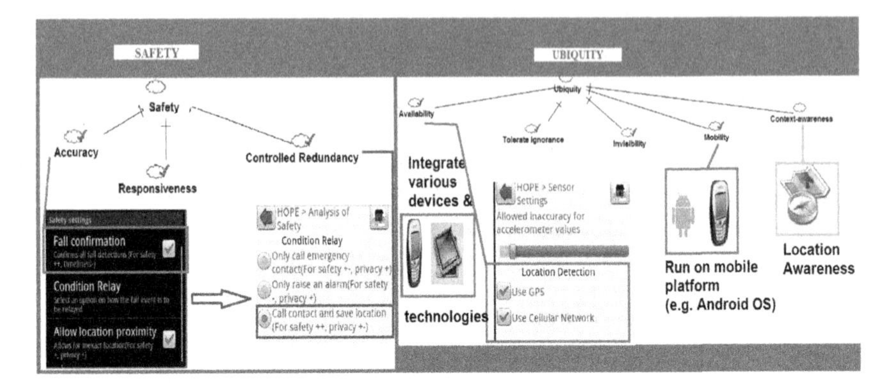

Fig. 9 Addressing ubiquity and safety softgoals in HOPE.

The emergency scenario addressed through the HOPE implementation is presented in Fig. 10.

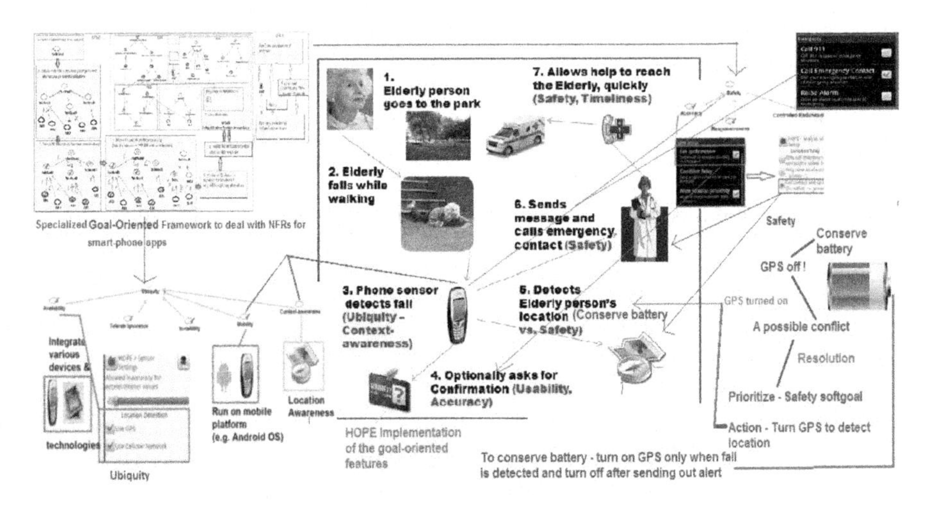

Fig. 10 Emergency scenario realization (A solution view).

4.3 User Interviews

The user evaluation is also an ongoing, iterative process and has been carried out 3 times so far. This evaluation is done using the currently running HOPE software system. This was carried out among 3 groups of people, a) Internal, among the research group students, b) Assistive persons such as nurses and other nursing home staff c) Real users (Seniors at a retirement facility). Of the three groups, the most important is c), with the real target users of the system. This experimentation was carried out among 7 senior participants at Grace Presbyterian Village, a well-known retirement/nursing home community in the Dallas area. The method used in obtaining user response, was one-to-one interviewing using the running HOPE system and the written user responses to the HOPE questionnaire.

The following are the details of the interview:

Number of participants: 7
Number of interviewers: 2
Time for one interview: 60 minutes
Type of interview: One-to-one, demonstration, questionnaire fill-up
Devices used: Motorola Atrix, Google G1
Average Age of Participants: >70

Some preliminary **results**:

- Features such as fall and location detection, user preferences aimed for ubiquity and safety NFRs were favored and termed easy to use.
- A separate icon for emergency help on all screens not favored by all.
- User-interface designed with usability NFR aspects such as minimum clicks, categorization favored among users.

4.4 Limitations

The HOPE system currently contains a limited set of features where goal alternative selection is carried out using our approach. As discussed in the section on trade-off analysis, there can be several possible conflicts that can occur, many that have not been safeguarded yet. We have begun work on identifying and dealing with design-time and run-time constraints for smart-phones, however, this being an ongoing work, still remains to be integrated with the HOPE application.

As far as fall detection is concerned, our focus has not been in implementing a rigorous fall detection algorithm considering parameters such as user height, weight, type of fall surface etc, thus, acknowledging the possibility of false positives. Additional sensors such as proximity or light have not yet been used.

Currently, there are limited mechanisms that provide flexibility to the user in altering/adding rules. A specific rule-based language has not been explored for implementation. There can be scenarios such as hardware or network faults where features may not function as expected even after applying our approach. We have not provided mechanisms, where in a situation, when no alternatives function– (e.g. GPS cannot receive signal, and there is no network connectivity, thereby there can be no location detection or call going through), we can add additional mechanisms dynamically to help satisfice the NFRs.

5 Related Work

The distinctive features of our work include dealing with non-functional requirements such as ubiquity and safety, especially for smart-phone applications, considering the smart-phone characteristics such as short lasting battery and built-in sensors. Our work shares the concepts found in some pieces of work [1, 8, 9, 10] in field of goal-oriented requirements engineering. We adopt and extend the NFR framework [1], in dealing with NFRs specifically for smart-phone applications. The goal-oriented approach is also found in [8], where the emphasis is more towards the functional aspects. In case of [9, 10], the emphasis is more towards the agent-goal dependencies, with a similarity of using a goal-oriented approach to deal with NFRs, differing in considering the characteristics of smart-phones or their applications in modeling a solution.

Our work shares the softgoal approach for ubiquity in work [11], where ubiquity is just shown to be composed of platform-independence and individualization, and [12], where ubiquity is shown as a softgoal affecting privacy and security. Our work goes beyond and extends them by further considering various alternative means as softgoals, while demonstrating our approach using a real application involving real users.

There have been efforts on developing ubiquitous sensor applications involving the elderly (See, for example, [13]). So far as automatic early fall detection of elderly persons is concerned, our work is similar to those in [14] and [15] in using common commercially available electronic devices to both detect the fall and alert authorities [15]. Our work additionally includes a proposal on taking a goal-oriented approach to taking an appropriate action whenever a fall is detected, based on the non-functional goals.

6 Conclusion

In this paper, we have proposed to treat non-functional requirements for smart-phone applications such as ubiquity, safety, as softgoals to be satisficed, whereby alternatives can systematically be explored and selected, while the various synergistic and antagonistic interactions among them are identified and analyzed. The approach deals with not only relationships among alternatives for satisficing a single NFR (e.g. intra-ubiquity analysis), but also with the relationships in the form of conflicts and synergies among various inter-related softgoals such as ubiquity and safety (e.g. inter-ubiquity-and-other-NFRs analysis). As a proof of concept, we have presented our smart-phone application, HOPE, using a goal-driven emergency scenario.

Future work includes the analysis of more inter-related softgoals, such as usability and cost, while adding conditions for safeguarding against undesirable side-effects. At the same time, we are working to strengthen our rule-base to accommodate more user-defined configuration changes. We have begun work on the dynamic/run-time goal prioritization/selection based on user preference/context. We also wish to conduct and analyze more user interviews based on further iterations of the HOPE system and the questionnaire.

References

1. Chung, L., Nixon, B.A., Yu, E., Mylopoulos, J.: Non-Functional Requirements in Software Engineering. Kluwer Academic Publishing (2000)
2. Weiser, M.: The Computer for the 21st Century. Scientific American (1991)
3. Lyytinen, K., Yoo, Y.: Issues and Challenges in Ubiquitous Computing. Communications of the ACM 45(12), 62–65 (2002)
4. Abowd, G., Mynatt, E.: Charting past, present, and future research in ubiquitous computing. ACM Transactions on Computer-Human Interaction 7(1), 29–58 (2000)
5. Krumm, J.: Ubiquitous Computing Fundamentals. CRC Press (2010)
6. Lutz, R.R.: Analyzing software requirement errors in safety-critical, embedded systems. In: Proceedings of the IEEE International Symposium on Requirements Engineering, pp. 126–133 (1993)
7. Wang, H., Mehta, R., Supakkul, S., Chung, L.: Rule-based context-aware adaptation using a goal-oriented ontology. In: Proceedings of the 2011 International Workshop on Situation Activity & Goal Awareness (SAGAware 2011), pp. 67–76 (2011)
8. Lamsweerde, A.: Goal-oriented requirements engineering: A guided tour. In: Proceedings of the 5th IEEE International Symposium on Requirements Engineering, pp. 249–262 (2001)
9. Yu, E.: Towards modeling and reasoning support for early phase requirements engineering. In: Proceedings of the 3rd IEEE International Symposium on Requirements Engineering, pp. 226–235 (1997)
10. Castro, J., Kolp, M., Mylopoulos, J.: Towards requirements-driven information systems engineering: the Tropos project. Information Systems, Elsevier 27(6), 365–389 (2002)
11. Akoumianakis, D., Pachoulakis, I.: Scenario Networks: Specifying User Interfaces with Extended Use Cases. In: Bozanis, P., Houstis, E.N. (eds.) PCI 2005. LNCS, vol. 3746, pp. 491–501. Springer, Heidelberg (2005)
12. Oladimeji, E.A., Chung, L., Jung, H., Kim, J.: Managing security and privacy in ubiquitous eHealth information interchange. In: Proceedings of the 5th International Conference on Ubiquitous Information Management and Communication, ICUIMC 2011 (2011)
13. Helal, S., Winkler, B., Lee, C., Kaddoura, Y., Ran, L., Giraldo, C., Kuchibhotla, S., Mann, W.: Enabling Location-Aware Pervasive Computing Applications for the Elderly. In: Proceedings of the 1st International Conference on Pervasive Computing and Communications, pp. 531–536 (2003)
14. Noury, N., Fleury, A., Rumeau, P., Bourke, A.K., Laighin, G.Ó., Rialle, V., Lundy, J.E.: Fall detection – Principles and Methods. In: Proceedings of the 29th Annual International Conference of the IEEE Eng. in Medicine and Biology Society, pp. 1663–1666 (2007)
15. Sposaro, F., Tyson, G.: iFall: An Android Application for Fall Monitoring and Response. In: Proceedings of the 29th Annual International Conference of the IEEE Eng. in Medicine and Biology Society, pp. 6119–6121 (2009)
16. The RE-Tools, `http://www.utdallas.edu/~supakkul/tools/RE-Tools`

Basic Object Distribution for Mobile Environment

Ji-Uoo Tak, Roger Y. Lee, and Haeng-Kon Kim

Abstract. Distributed objects usually refers to software modules that are designed to work together, but reside either in multiple computers connected via a network or in different processes inside the same computer. One object sends a message to another object in a remote machine or process to perform some task. Distributed objects are a potentially powerful tool that has only become broadly available for developers at large in the past few years. The power of distributing objects is not in the fact that a bunch of objects are scattered across the network. This paper presents an environment for supporting distributed application using shared object in Java within a heterogeneous environment for mobile applications. In particular, It offers a set of objects such as Lists, Queues, Stacks that can be shared across a network of heterogeneous machine in the same way as DSM systems. Shared is achieved without recourse to Java RMI or object proxies as in other object systems. An implementation of the environment MBO(Mobile Business Objects) is provided together with performance timings.

Keywords: MBO(Mobile Business Objects),Java RMI, Distributed Object, Distributed Systems, Shared Distributed Memory.

Ji-Uoo Tak
Department of Computer Engineering, Catholic University of Daegu, Korea
e-mail: lebbenle@cu.ac.kr

Roger Y. Lee
Software Engineering & Information Technology Institute, Central Michigan University, USA
e-mail: leelry@cmich.edu

Haeng-Kon Kim
Department of Computer Engineering, Catholic University of Daegu, Korea
e-mail: hangkon@cu.ac.kr
Corresponding author.

R. Lee (Ed.): Software Engineering Research, Management and Appl. 2012, SCI 430, pp. 127–140.
springerlink.com

1 Introduction

Making software a commodity by developing an industry of reusable components was set as a goal in the early days of software engineering. While significant progress has been made, this still remains a long term challenge. The function of middleware for mobile computing is to mediate interaction between the parts of an application, or between applications. Therefore architectural issues play a central role in middleware design. Architecture is concerned with the organization, overall structure, and communication patterns, both for applications and for middleware itself. Besides architectural aspects, the main problems of middleware design are those pertaining to various aspects of distributed systems. Any middleware system relies on a communication layer that allows its different pieces to interoperate. In addition, communication is a function provided by middleware itself to applications, in which the communicating entities may take on different roles such as client-server or peer to peer. Middleware allows different interaction modes (synchronous invocations, asynchronous message passing, coordination through shared objects) embodied in different patterns.

The designers of future middleware systems face several challenges:

- Middleware systems rely on interception and indirection mechanisms, which induce performance penalties. Adaptable middleware introduces additional indirections, which make the situation even worse.

- As applications become more and more interconnected and interdependent, the number of objects, users and devices tends to increase. This poses the problem of the scalability of the communication and object management algorithms, and increases the complexity of administration.

- Ubiquitous computing is a vision of the near future, in which an increasing number of devices embedded in various physical objects will be participating in a global information network. Mobility and dynamic reconfiguration will be dominant features, requiring permanent adaptation of the applications.

- Managing large applications that are heterogeneous, widely distributed and in permanent evolution raises many questions, such as consistent observation, security, tradeoffs between autonomy and interdependence for the different subsystems, definition and implementation of resource management policies.

The object technology has proven to be a suitable paradigm for programming distributed systems. Several systems have been devised. At present, the technologies of objects and distributed programming are unified giving rise to the distributed object technology (DOT) [7]. The web technology is also integrated with the DOT. One good example is Java [2] of Sun Microsystems. Java is a an object oriented language that give support to distributed programming through the concept of distributed objects. Characteristics of Java include: portability of Java code on different platforms (e.g. Unix, Windows, Mac), support of web programming. Sharing in Distributed systems is one essential characteristic, but it is a difficult task to achieve. Several DSM systems have been provided [8] but are limited and complex. One interesting issue is to investigate sharing across different platforms, and in particular in a transparent form is the usage of the concept of an object. For this purpose, we suggest a library of shared object that can be accessed across any node of a heterogeneous system in mobile computing environments. We presents an environment for supporting distributed application using shared object in Java within a heterogeneous environment for mobile applications. In particular, It offers a set of objects such as Lists, Queues, Stacks that can be shared across a network of heterogeneous machine in the same way as DSM systems. Shared is achieved without recourse to Java RMI or object proxies as in other object systems. An implementation of the environment MBO is provided together with performance timings.

2 Background

2.1 Distributed Systems and Sharing

The progress in the technology of VLSI design, cheap processors, and high speed network made it possible to interconnect several machines to form very powerful computer systems. These machines have the objective of supporting applications which consume a lot of processing power as in the case of visualization applications, CAD/CAM, Distributed database, etc [9]. There are two types of systems: multiprocessors and multicomputers. Multiprocessor systems exhibit a semantic of previsible performance, but have a very complex hardware and difficult to construct. Programming these machines is simple; the programs share a common share a common address space. Multicomputer systems consists of a set of processors; each with its own memory and is connected to another through a high speed network. These systems are easy to build physically but the software is more difficult since it requires the programmer structuring its application as intercommunicating processes. With the intention of combining characteristics of multiprocessors and mulicomputers, DSM (Distributed Shared Memory) systems appeared. These provide a global virtual memory accessed by all processors. However, the software is difficult to build and has an impermissible performance [7].

2.2 Object and Distribution

The object oriented technology has been pointed to be a promising one to control the complexity generated by distributed systems. The concepts of object orientation such as modularity, access to data through the interface make the use of objects appropriate to model distributed systems. This approach has been enforced and well accepted with the development of several distributed environments that use the concept of objects (e.g. CORBA of OMG[12], DCOM and OLE of Microsoft. Another aspect that should be taken into consideration is that object oriented applications are

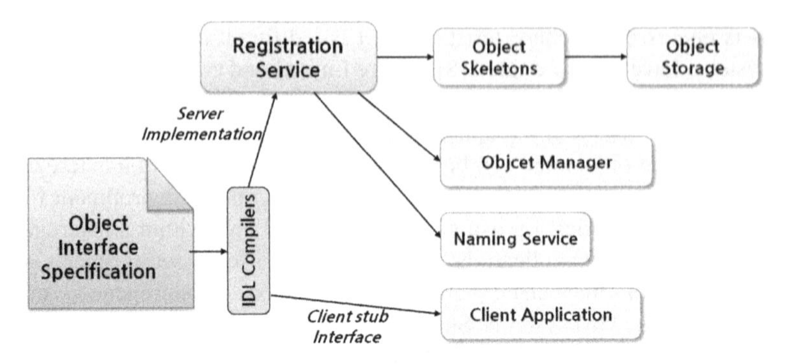

Fig. 1 General architecture for distributed object systems

programmed in terms of communicating objects which is a natural form to program distributed systems [7,10]. The technology of distributed objects (DOT) is revolutionizing the programming methodology in distributed systems. In fact it includes three technologies of: objects, distribution and Web. The object technology (OT) was introduced in the seventies by Adele Goldberg and Alan Kay in a language called Smalltalk. In thin model, objects encapsulate structures and behaviour; each is an instance of a class that belong to a class hierarchy. The OT has matured quickly and became well accepted as a technology that encapsulates complexities, improves maintenance, promotes reuse and reduces the cost of life cycle of software. The focus of DoT has been to do the atmosphere of more transparent computation with regard to the use of computation engines and computation objects. With the Web technology (WT), born in the nineties, it caused a quick explosion in the use of the Internet. In 1995, Java came and gave more initiative for the usage of DOT. The key concept behind DOT is the interconnection. In currently, many abstraction levels can be used to describe the connection between mobile and net [7]. DOT brought the concept of objects with the notion of transparency of the distributed computation [10]. Objects became a skilled technology for distributed processing.

2.3 Distributed Systems

A distributed system is simply one or more processors that do not share memory or a clock and communicate via some means to accomplish some task. The processors (nodes) in a distributed system can be almost anything that we are familiar with today, workstations, personal computers, mainframes, hand-held devices, cellular telephones, etc. Communication between nodes can be handled in a number of ways depending on the system. Some of the most common methods of communication are the WAN (wide-area network), LAN (local-area network), and high-speed bus. There are four advantages to a distributed system, resource sharing, reliability, computation speedup, and communication. In a distributed system, each node can share another nodes resources (depending on topology). The sharing of resources is also a reason why a distributed system can be more reliable (again, depending on topology). Unlike an autonomous workstation or personal computer where if one resource fails the system fails, in a distributed system the failure of one node does not cause the failure of all of the nodes; thus the system should continue to operate even after the loss of some resources. Because a distributed system is designed to share resources, it can naturally be designed to share computation as well. This is where users on a distributed system have an opportunity to share their own resources with other users. A good example of this is email, where the users resources are their thoughts in the body or subject of the message; other examples are the sharing of files, or serving a website. There are several types of distributed computing models (or paradigms) as in figure 2. The two most popular paradigms today are peer-to-peer and client-server. In a peer-to-peer setting, each node has the option to be in direct communication with every other node in the system. In a client-server paradigm, a client node will only communicate one other server node with will then retrieve the resources the client is requesting. For most mobile and ubiquitous applications, these paradigms seem to be the best suited because of their speed, reliability and adaptability.

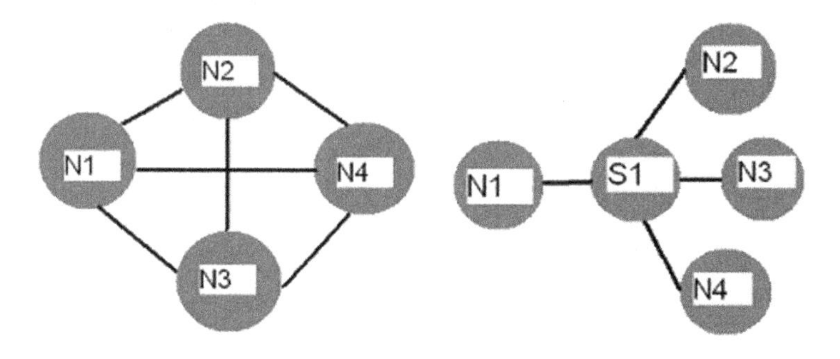

Fig. 2 Distributed Systems Peer-To-Peer and Client-Server Distributed Paradigm

2.4 *Java RMI*

Java is not just a language, but it has an entire philosophy of advanced software engineering. The Java system presents several advantages, because it is simple, object oriented, robust, portable and architecture neutral; it adopts modern methodologies (client/server), interpreted, besides it is multithread. One of the most interesting characteristics of Java is its portability. This facilitate the development of applications for heterogeneous environments in a network. This implies that Java application can run on different hardware architectures without need of fittings. Instead of creating an executable code for a certain machine, Java generates bytecodes, that are interpreted and executed by a virtual machine (neutral architecture), that is written for several types of platforms. Although Java is interpreted, it may present good performance, since the creation of the code does not have the cycle of compile-link-load-test-crash-debug of other languages; now it is simply complied and run.

One in the most usual ways of communication among processes in a distributed system is the remote procedures call (RPC). Java is one language that uses this mechanism and uses RMI (Remote Method Invocation). Figure 1 illustrates the RMI system. The RMI system is divided into four parts :

- Stub/skeleton: the stubs are used by clients (proxies) and the skeletons are used for the servers (dispatchers);
- Remote reference: reference behaviour and invocation (using unicast or multi-cast);
- Transport: configuration and management of the connection;
- Distributed garbage collection: reference of remote objects.

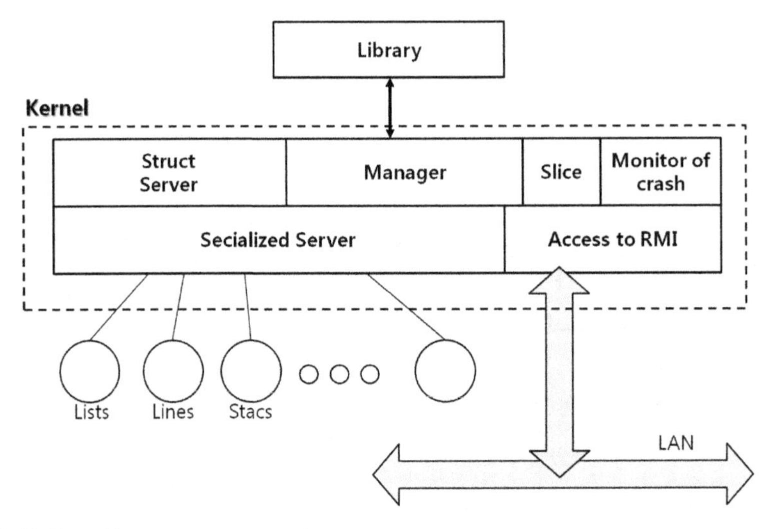

Fig. 3 RMI Architecture

Java supports multithreading. Monitors are used to prevent inconsistencies o data due to multiple accesses which may arise. According to Andrews [1], a monitor is a mechanism of abstraction of data: they encapsulate the representations of the resources and they supply procedures that act in these resources. The mutual exclusion is guaranteed since a procedure is just executed one at a time (similar CCR) and conditional synchronization is obtained through conditional variables.

3 The Model of Our Environment

Our environment provides a collection of objects (data structures) that can be shared across a heterogeneous network. It has the following characteristics:

- Access to elements of objects (data structure elements) are done through methods implemented within objects. For example, insert to put an element to an object, and delete to remove an element from the object.
- There is no difference in accessing a local or remote object.
- Provides method for creating objects: local or remote.
- Objects are shared between nodes of the network.

The proposed environment is illustrated in Figure 2. The environment is presented with three heterogeneous computers; however, the system may have any number of nodes. The architecture has the following components:

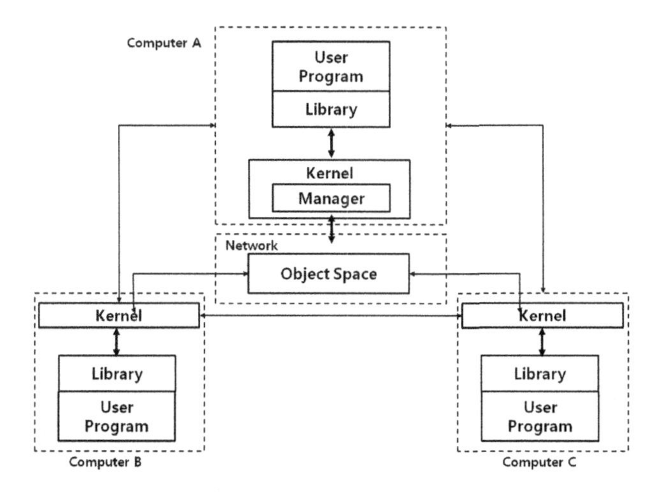

Fig. 4 Architecture of our Environment

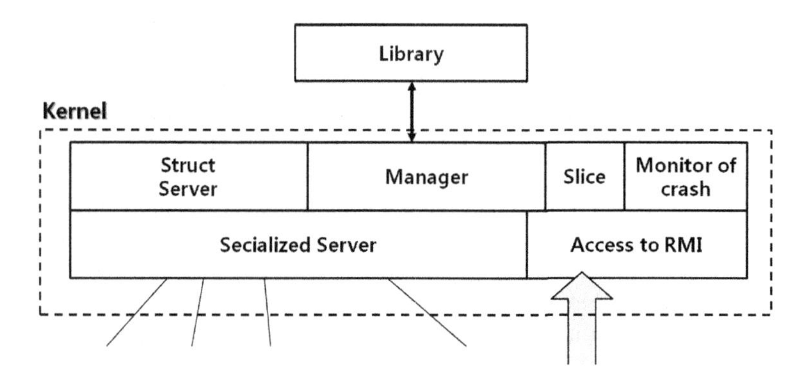

Fig. 5 kernel without manager

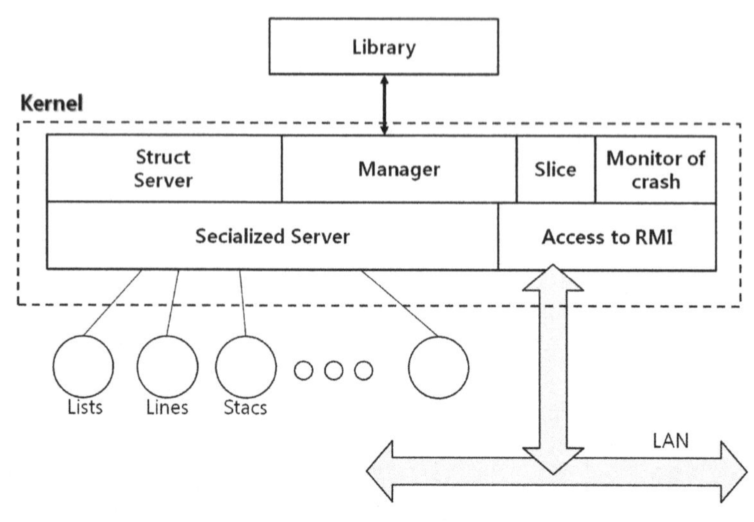

Fig. 6 Kernel with Manager

Object space: Each node has its own object space or local repository where the objects are stored. The sum of all repositories is called the global repository which is managed by the manager component. It is implemented by the language Java through RMI.

Kernel: Each node has its own kernel. The role of the kernel is to keep track the allocation of data structures. The kernel provides functions to create and remove objects. It also allows other kernels to access local data structures in a safe and efficient form. A kernel consists of sub components: server structures, specialized servers, slice and RMI access. The server structure is responsible for creating and deleting the structures. The specialized server (e.g. List server, stack server, queue server) are responsible for servicing specific object requests. The slice component is responsible to compute the load in each node every five minutes. The load is used by the manager to best distribute objects over the network. There are two types of kernels: a kernel with a manager figure 3-a and a kernel without a manager figure 3-b. In a network there is only one kernel with a manager all of other nodes consist a one kernel without a manager. The manager is responsible for decision such as where an object should be created. This decision is taken depending on the load of each node. For this purpose, the manager maintains a table of information containing a load of each node, estimate time slice of each process, amount of memory available, machines addresses (Ips). The crash monitor component verifies which machine has crashed. If it happens, the manager is called to take appropriate action such as taking the machine out of the system.

Library: provides a set of functions (methods) for manipulating the objects (i.e. data structures).The functions are provided as part of the interface of each object.

There are several types of objects:

Lists: Lists can be created in three forms: local, remote and specified location. A local list is created on the same node where the create method is invoked. The remote list is created on a remote node chosen by the system. A list created in a specified location is created on a node specified by the user (IP specified).

Stacks: three types of tacks may be created as well in the same manner as object Lists.However, the Stacks are accessed in LIFO through operations push and pop.

Queues: Objects queue may be local remote or location specified. Accesses to such object is in FIFO through operation of insert and remove.

3.1 Manipulating Objects

The usage of the library is very simple. Creating and invoking methods on shared object can be done in the same way as in Java. A shared object may be created as local, remote or with specified location. Constructors are provided for appropriate types of objects. List-local(Name) for initialising a local object with Name, List-remote(Name) to initialise a remote object with Name and List-remote(Name, IP) to initialise a remote object specified at the node IP with Name. A programmer may create a remote object and initialise it as follows:

List-remote list – new List-remote (hello);

This has the effect of creating a remote object list of type List and initialise it with an element hello. Manipulating an object is also simple. For example, inserting an element world to the created object may be done as follows:

list.insert(world); // insert the an element world to the object List list.

As it can be seen, programming with shared object is very simple. The programmer does not need to get involved with details of communication, RMI, object localization etc.

4 Evaluation of the Proposed Environment

A distributed application allows an arbitrary number of running components (programs) across any number of address spaces and nodes, cooperating as peers in implementing the application. Components of an *object-based* distributed application communicate and cooperate through *shared objects.* Shared objects are fine-grained, fully encapsulated passive entities that consist of private internal state and a set of operations by which that state can be accessed and modified. Figure 7 shows the overall UML sequence diagram for MBO behavior.

The design cornerstone of the proposed system platform is the Mobile Business Object (MBO). The MBO encapsulates the business process logic and data into a reusable unit that can be accessed from a variety of clients. The development can take two approaches, firstly the Top-Down approach, which creates the MBO based upon how the client wants to use it eventually binding to the data source. Secondly, the Bottom-UP approach which uses the data source to automatically bind and create the MBO. Both use the tooling and are easy to create, with similar steps listed below :

1. Connect to the back-end data sources that need to be mobilized, via the Unwired tooling
2. Connect the Unwired tooling to the Unwired Server
3. Create a Mobile Application Project
4. Create the MBO from the data source via wizards or Create the MBO attributes and Create, Update, Delete Operations separately
5. Deploy the MBO's to the Unwired Server

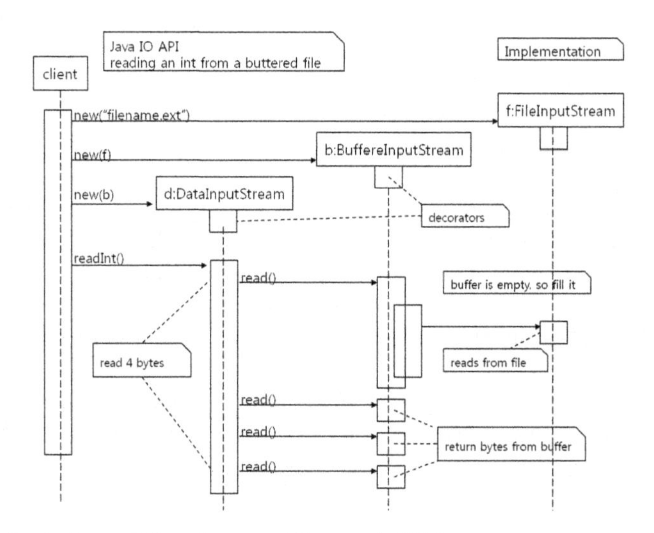

Fig. 7 Mobile Business Object Sequence Diagram in UML

Diving a bit deeper the MBO is

- A metadata definition representing a slice of enterprise data and operations
- A design-time element that manifests as a concrete runtime object
- The building block for synchronization and transaction of multiple elements of an application

- Personalization keys (device context) helps narrow data sets *see below
- MBOs are the basis for concrete Objective-C, C# and Java device-side object interfaces

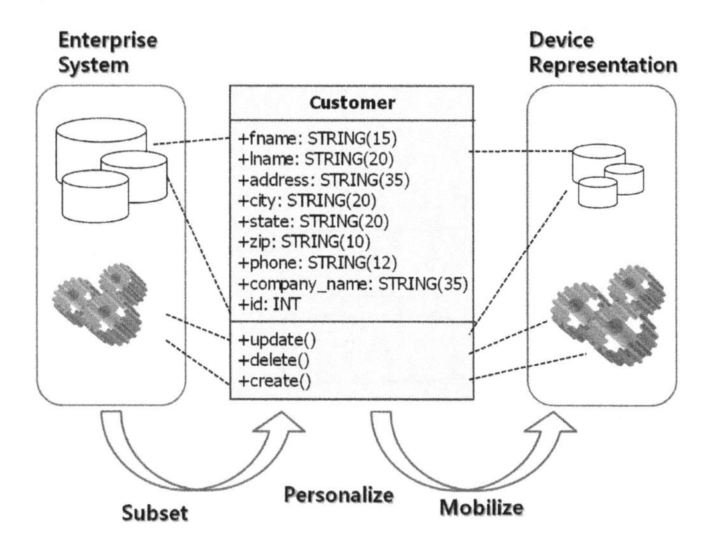

Fig. 8 Relationship between Enterprise System and Mobile Device

Attributes (synchronized data)

- Define the scope of the device-side data store
- Attributes and parameters in an MBO definition form the shape of the server-side cache
- Server cache and device is populated by reading data from the enterprise, e.g., through an MBO definition operation like a SQL select statement

Operations (enterprise playback)

- Backend tasks to be carried out within the enterprise or through an external service, typically with respect to the MBO attributes

Parameters

- Can be passed to one or more operations
- Linked to upstream parameters or attributes
- Defined by personalization keys for a user or group *see below

Relationships

- Form the basis of transactions by identifying the dependencies and state requirements of device-side entities. Relationships are formed by mapping attributes and parameters similar to a primary-foreign keys :

1. Design or Modify in the Designer Mode through the GUI interface to define/Edit Mobile Business Objects Use tooling set preferred (Eclipse or Visual Studio)?
2. In designer mode, the SUP tool generates the SQL and code operations for the devices and server side apps and any addition API, filtering and so forth occurs in this phase. Then package and deploy.
3. In operation or runtime mode, SUP Runtime engine then performs the synchronization/loads and relationship management as specified by the MBO runtime parameters

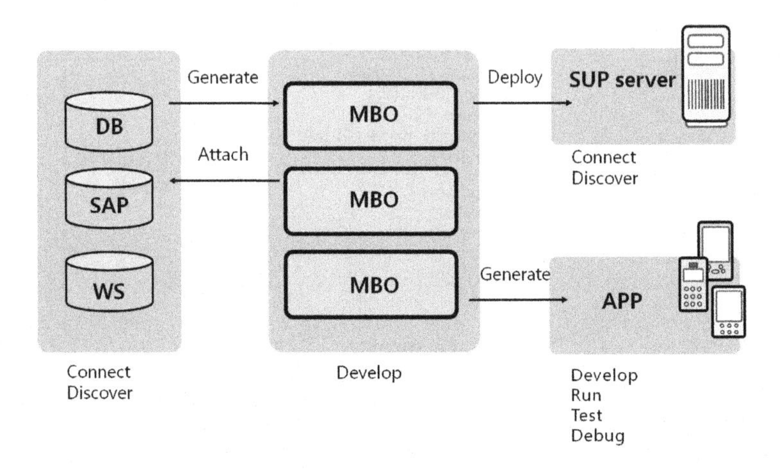

Fig. 9 Architecture for MBO in this paper

AS figure 9, Our Platform offers flexibility when creating mobile applications ? catering to the needs/requirements of both novice and advanced mobile developers. We used the Bottom-Up approach and the wizards in the Unwired Tooling to create a 1:1 representation of the tables that I wanted to synchronize onto Mobile Client.

5 Conclusions

In a distributed system, processes and users require access to shared data for meaningful cooperation. Traditional high performance computing often uses message passing to share information in parallel applications on distributed environments. With this approach, the developer has the advantage to control communication occurring in these applications and can adjust it to avoid unnecessary latency which can effect overall performance. This control, however, implies a responsibility to plan every communication detail. As a result, application development becomes extremely difficult and time consuming. In this paper, we have presented an environment for supporting distributed applications within a heterogeneous network of machines for mobile computing, In particular, if offers a set of shared objects that

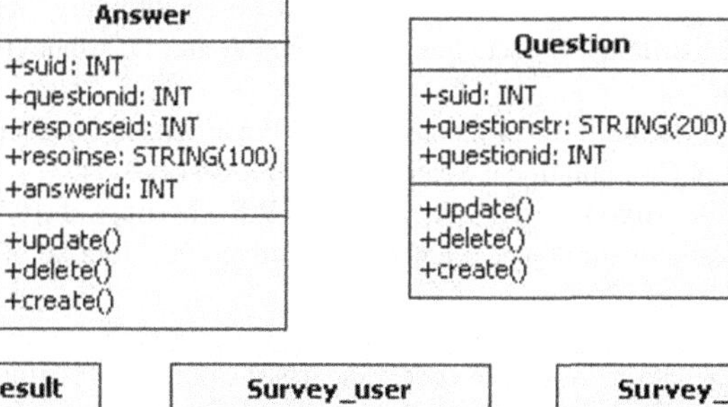

Fig. 10 Classification for Mobile Object Distribution

can be accessed from any machine in the same way as any other object without recourse to RIM or object proxies. This will facilitate the programmers task. An implementation using a centralized model has been presented together with some timings. The results can be improved if a decentralized strategy is used. As part of future development, we are in a process of distributing the object physically.

Acknowledgments. This work was supported by the Korea National Research Foundation (NRF) granted funded by the Korea Government (Scientist of Regional University No. 2012-0004489).

References

1. Akinnuwesi, B.A., Uzoka, F.-M.E., Olabiyisi, S.O., Omidiora, E.O.: A Framework for user-centric model for evaluation the performance of distributed software system architecture. Expert Systems with Applications (in press, 2012)
2. Patil, S., Kobsa, A., John, A., Seligmann, D.: Methodological reflections on a field study of a globally distributed software project. Information and Software Technology 53(9), 1568–1578 (2010)
3. Prikladnicki, R., Audy, J.L.N.: Process models in the practive of distributed software development: A systematic review of the literature. Information and Software Technology 52(8), 779–791 (2010)
4. Romanazzi, G., Jimackb, P.K., Goodyer, C.E.: Reliable performance prediction for multigrid software on distributed memory systems. Advances in Engineering Software 42(5), 247–258 (2011)

5. Hu, Z.-H., Ding, Y.-S.: An immune inspired co-evolutionary affinity network for prefetching of distributed object. Journal of Parallel and Distributed Computing 70(2), 92–100 (2010)
6. Pendharkar, P.C.: A multi-agent memetic algorithm approach for distributed object allocation. Journal of Computational Science 2(4), 353–364 (2011)
7. Ferreira, K.A.M., Bigonha, M.A.S., Bigonha, R.S., Mendes, L.F.O., Almeida, H.C.: Identifying thresholds for object-oriented software metrics. Journal of Systems and Software 85(2), 244–257 (2012)
8. Olszak, A., Jłrgensen, B.N.: Remodularizing Java programs for improved locality of feature implementations in source code. Science of Computer Programming 77(3), 131–151 (2012)
9. Wang, J., Jiang, Y.: An Improved Algorithm About Failure Agreement Based on Distributed Shared Memory. Procedia Engineering 15, 3330–3334 (2011)
10. Chiu, Y.-C., Shieh, C.-K., Huang, T.-C., Liang, T.-Y., Chu, K.-C.: Data race avoidance and replay scheme for developing and debugging parallel programs on distributed shared memory systems. Parallel Computing 37(1), 11–25 (2011)
11. Huang, G., Wang, W., Liu, T., Mei, H.: Simulation-based analysis of middleware service impact on system reliability: Experiment on Java application server. Journal of Systems and Software 84(7), 1160–1170 (2011)

Enhancing Tool Support for Situational Engineering of Agile Methodologies in Eclipse

Zahra Shakeri Hossein Abad, Anahita Alipour, and Raman Ramsin

Abstract. In recent years, with the growth of software engineering, agile software development methodologies have also grown substantially, replacing plan-driven approaches in many areas. Although prominent agile methodologies are in wide use today, there is no method which is suitable for all situations. It has therefore become essential to apply Situational Method Engineering (SME) approaches to produce agile methodologies that are tailored to fit specific software development situations. Since SME is a complex process, and there is a vast pool of techniques, practices, activities, and processes available for composing agile methodologies, tool support–in the form of Computer Aided Method Engineering (CAME) environments–has become essential. Despite the importance of tool support for developing agile methodologies, available CAME environments do not fully support all the steps of method construction, and the need remains for a comprehensive environment. The Eclipse Process Framework Composer (EPFC) is an open-source situational method engineering tool platform, which provides an extensible platform for assembly-based method engineering in Eclipse. EPFC is fully extensible through provision of facilities for adding new method plug-ins, method packages, and libraries. The authors propose a plug-in for EPFC which enables method engineers to construct agile methodologies through an assembly-based SME approach. The plug-in provides facilities for the specification of the characteristics of a given project, selection of suitable agile process components from the method repository, and the final assembly of the selected method chunks, while providing a set of guidelines throughout the assembly process.

Keywords: Agile methodology, Situational Method Engineering, Eclipse Process Framework Composer, Computer-Aided Method Engineering.

Zahra Shakeri Hossein Abad · Anahita Alipour · Raman Ramsin
Department of Computer Engineering, Sharif University of Technology, Tehran, Iran
e-mail: z_shakeri@ce.sharif.edu, alipour@ce.sharif.edu, ramsin@sharif.edu

R. Lee (Ed.): Software Engineering Research, Management and Appl. 2012, SCI 430, pp. 141–152.
springerlink.com

1 Introduction

The simplicity and development speed of agile methodologies are the main reasons for their popularity. Although prominent agile methodologies are available, there is no general-purpose agile methodology which fits all situations. This has led to the application of Situational Method Engineering (SME) approaches to produce project-specific methodologies that are tailored to fit specific development situations. Like all engineering disciplines, efficient application of SME methods is dependent on the availability of adequate tools; Computer Aided Method Engineering (CAME) environments have been developed for this purpose [1, 2].

There are three main SME approaches [3]: *Assembly-based* SME, in which a method is constructed from reusable method components that are extracted from existing methodologies and stored in a repository called the "method base"; *Extension-based* SME, in which existing methods are extended and enriched by applying extension patterns [1]; and *Paradigm-based* SME, in which a new method is constructed by instantiating a metamodel or applying abstraction to existing methods. Among the different approaches to SME, *Assembly-based* SME is the most commonly used and has become the basis of method construction in CAME tools. Method development in these tools consists of three distinct stages: Specifying the method requirements based on the situation of the project, selecting the appropriate method fragments, and assembling the fragments into a coherent methodology.

In assembly-based engineering of agile methodologies, CAME tools are expected to provide the necessary means for performing the following four stages [2]: (1) Specification of a set of methodology requirements by characterizing the project at hand; (2) Development of an agile method base (repository) by extracting a set of method fragments from existing agile methodologies; (3) Matching the extracted method chunks with the methodology requirements, thereby forming a cohesive set of method chunks; and (4) Supplementing the method chunks with guidelines on how they can be assembled into a coherent process. The main shortcoming of existing CAME tools is that they only partially cover these stages [1].

The Eclipse Process Framework Composer (EPFC) [4] is the single most prominent CAME tool currently used by method engineers. EPFC already provides support for the instantiation of XP and Scrum methodologies, but this support is partial; EPFC represents these methodologies as general methods, and does not support assembly-based SME stages for constructing bespoke methodologies from their components. In order to address the shortcomings of existing CAME tools, we propose ASEAME (Assembly-based Situational Engineering of Agile Methodologies in Eclipse) as a plug-in for EPFC which enables method engineers to construct agile methodologies through an assembly-based SME approach. ASEAME provides facilities for the specification of the characteristics of a given-project, selection of suitable agile process components from the method repository, and the final assembly of the selected method chunks, while providing a set of guidelines throughout the assembly process.

The rest of this paper is organized as follows: Section 2 briefly reviews the literature related to this work and highlights the contributions of this research; Section 3 explains the details of ASEAME; Section 4 evaluates ASEAME according to the

ISO/IEC 9126 quality model and compares it to other CAME tools; and Section 5 presents the concluding remarks and suggests ways for furthering this research.

2 Related Research

Since the early years of method engineering, several academic prototypes of CAME environments have been introduced [1], and different versions of them have been developed. Since our main focus is on CAME tool support for *agile* methodologies, we have divided this research work into the following categories:

- *Research conducted on CAME environments in general:* None of the research efforts in this category has resulted in CAME tools that provide adequate support for the method engineering process. Method Base [5] is one of the primary tools in this area, which is focused on helping the method engineer in selecting the appropriate method for the project at hand. This tool does not support some of the features of assembly-based SME, but provides facilities for method customization. Other CAME tools in this category, including Decomerone [1], MENTOR [6], Method Editor [7], MERU [8], and METAEdit+ [9] provide partial support for the assembly-based approach, but only MENTOR and MERU cover method requirements analysis; others just support the design and implementation stages. MENTOR and MERU, in turn, have other problems: In MERU, methods are considered only in the product part; therefore, method fragments defined in this tool are not comprehensive; since method fragments collection is a prerequisite for method fragments selection, we regard this is as a major flaw. MENTOR, on the other hand, supports both assembly-based and paradigm-based approaches, and the assembly-based approach is not its main focus.

- *Research conducted on CAME-tool support for agile methodologies*: Among the few studies that have been conducted in this area, EPFC is the only tool introduced that represents two agile methodologies (SCRUM and XP) in their entirety, and that provides the means to instantiate these methodologies. However, this cannot be considered as adequate support for assembly-based SME [2]. In [10], a toolbox is introduced for agile methodology selection which assists the method engineers in classifying the projects, selecting agile methodologies, and selecting agile practices. However, the method classification factors provided in this tool are very limited. Moreover, the tool proposes a limited collection of agile practices for the project at hand, and even these limited practices are provided separately for each agile methodology. Furthermore, the ultimate assembly of the practices is not supported; in other words, this toolbox does not support the assembly-based approach.

3 Particulars of ASEAME

ASEAME is an Eclipse plug-in for EPFC which supports the comprehensive implementation of assembly-based SME, and enhances situational method engineering of agile methodologies. As shown in Fig. 1, EPFC provides extension facilities for adding new method plug-ins, method packages, and method libraries [1]. In this section, we first study the coverage of the generic SME process in EPFC, and then introduce the architecture of ASEAME.

3.1 Coverage of the Generic SME Process in EPFC

Although EPFC is being used extensively by methodology engineers for ME and SME purposes, this environment does not provide full coverage of SME stages, and requires a high level of involvement by the method engineer. In this section, we investigate the mapping between the generic SME stages (as defined in [4]) and the EPFC, and present ASEAME with the purpose of providing complete coverage of SME stages, so that the deficiencies of EPFC are properly addressed.

EPFC is an open-source Eclipse project which has been created using the Eclipse Integrated Development Environment, and which supports a large number of Eclipse plug-ins. This CAME environment enables process engineers and managers to implement, deploy, and maintain situation-specific methodologies, and is intended to provide the following two facilities:

- A knowledge base to help developers learn their responsibilities in SME projects. This knowledge base includes external content as well as the users' own content, and can be used for educational purposes.
- A catalog of predefined processes which helps method engineers learn how to perform their responsibilities in a process, and understand how the different tasks in a process relate to one another. Some of these processes are complete "deliverables" that can be adapted to individual situations. Other processes (called "capability patterns") are building blocks for other complete processes, and represent the best development practices for specific disciplines, technologies, or management styles.

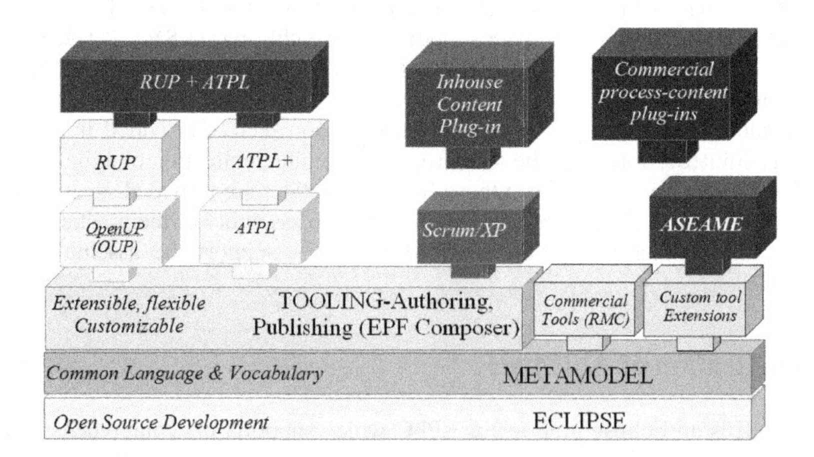

Fig. 1 Position of ASEAME in the EPFC architecture

In [4], a generic process is proposed for SME which consists of three main steps:

- *Situation characterization:* In which project situations are distinguished by using a set of factors called "situational factors." Project managers and method engineers can specify the situation of the project at hand by assigning values to these factors.
- *Method fragments selection:* In which the method fragments that correspond to the project characterization are selected from the method base (repository).
- *Method assembly:* In which the selected method fragments are assembled to form a coherent situational method. Method engineers need proper guidelines and rules to develop consistent methods.

EPFC does not support the "situation characterization" stage. It only supports two stages: Selection of method fragments, and method assembly. However, this support is only partial, and requires a high level of involvement on the part of the method engineer. Since characterizing the situation of the project at hand is a key step in determining method requirements, ignoring this step reduces the quality and efficiency of the methodologies produced.

Considering the vast amount of activities, tasks, and techniques available for agile methodologies, the method engineer's deep involvement in fragments selection not only significantly increases the complexity of the task, but also turns it into an error-prone process. Moreover, if method engineers do not have sufficient knowledge of all agile methodologies, necessary and useful method fragments may be ignored. ASEAME is specifically intended to address these issues.

3.2 ASEAME Architecture

EPFC uses the System Process Engineering Metamodel (SPEM 2.0) standard for method decomposition [17]. In SPEM 2.0, method content is made up of reusable components that compose processes. Elements are of three types: roles, work products, and tasks. To organize the components and define them at different levels of abstraction, and also to delimit the sequence of the activities performed, SPEM 2.0 incorporates the concepts of lifecycle, phase, activity, task, and technique (in descending order of granularity). As mentioned in [2], we have adopted SPEM 2.0 in defining the proposed agile method base (used in ASEAME's method repository).

Throughout the rest of this section, we present a complete description of ASEAME, based on the stages of assembly-based SME that it addresses. Fig. 2 and Fig. 3 show the ASEAME screens corresponding to these stages.

3.2.1 Situation Characterization

ASEAME characterizes the project at hand through defining a series of situational factors for agile methodologies. As shown in Table 1, these factors are organized in three groups: Application Domain, Project Organization, and Environment.

A default value is defined for each group. Discussion on how to select these factors, however, is out of the scope of this paper. The interested reader is referred to [1, 18, 19, 20].

After these factors have been initialized, the method requirements will be determined, and the input for the next step (method fragments selection) is provided.

Table 1 Situational Factors for Agile Methodologies

	Decision Factors	Possible Values in ASEAME
Environment	Degree of financial constraints	Normal / High
	Diversity of end-users	Wide / Narrow
	Time pressure imposed on the project	Yes / No
	Degree of importance of the project to the environment	Yes / No
Application Domain	Degree of formalism required in the methodology	Low / High
	Criticality of methodology quality factors	Normal / High
	Size of the target system	Normal / Large
	Criticality level of the target system	Normal / High
	Technology innovation level of the target system	Normal / High
	User-interface dependency of the target system	Low / High
Project Organization	Degree of developers' business knowledge	Adequate/ Inadequate
	Degree of developers' technical expertise	Adequate/ Inadequate
	Geographical distribution of development teams	Yes / No
	Distribution of skills among teams and members	Even / Uneven
	Degree of teams' acquaintance with agile methodologies	Adequate/ Inadequate

3.2.2 Method Fragments Selection

Once the situational factors are initialized by the method engineer, ASEAME provides facilities for selecting the appropriate method fragments.

Several approaches exist for method fragments selection: The Map [21] approach selects method fragments by measuring the similarity between the requirement's map and method fragments. However, such calculations may not provide sufficient distinction between method fragments, and the selected fragments might be similar. Therefore, choosing the appropriate method fragments may require a higher degree of involvement by the method engineer [20].

An alternative approach uses the Deontic matrix [22] for method fragments selection. This matrix is a two dimensional array of process elements, spanning activities, tasks, and techniques. These matrices can be developed at different process component levels, such as task/activity, task/technique, activity/technique, and work-product/activity. Cell values specify the connection among these process components. The matrix can easily become huge if a large number of components are involved; filling the matrix can therefore become a time-consuming task for the method engineer.

The activity diagram approach [23, 24, 25] uses UML activity diagrams for selecting method fragments. Due to its low level of formalism, and the fact that diagram development heavily depends on the knowledge of the designer, implementing this approach in CAME-tools has proved difficult [23].

A multicriteria approach is also available [20], which resembles a modified version of the original assembly approach. Multicriteria techniques are commonly used in decision making for determining priorities based on the available alternatives. This is the approach used in ASEAME for method fragments selection.

3.2.3 Method Assembly

In [26], Brinkkemper presents techniques for assembling method fragments at both product and process levels. This approach formalizes the assembly process by defining rules and guidelines. In [5], Harmsen uses a set of rules in the form of mathematical axioms and derived corollaries and theorems for assembling method fragments. These rules focus on situation-independent factors such as completeness, consistency, efficiency, soundness, and applicability. Learning and implementing these rules can be very time-consuming for the method engineer. In addition, correct assembly is dependent on the method engineer's knowledge of mathematical formalisms, which might not be adequate. ASEAME supports this step, which has a strong impact on method consistency, through examining the selected method fragments from different aspects. ASEAME also provides guidelines in order to resolve the conflicts arising among rules, and also to address nonfunctional requirements in the final method.

3.3 Examples of ASEAME Screens

In this section, examples of ASEAME screens are shown. Fig. 2 shows the screen on which the method engineer will specify the *organization situational factors* of the project. ASEAME provides screens for entering other types of situational factors, including *environment situational factors* and *application domain situational factors.* Fig. 3 shows an example of the final results of ASEAME, consisting of assembly guidelines and a coherent agile methodology composed of phases, activities, tasks, techniques, roles, and work products.

4 Analysis of the Proposed Plug-In

The ISO/IEC 9126 quality model [27] will be used in this section for evaluating ASEAME. This model evaluates software systems by inspecting six main features. Considering that this model spans a huge number of standards, not all of which are applicable to CAME tools, an adaptation of ISO/IEC 9126 (introduced in [1]) is used for evaluating ASEAME.

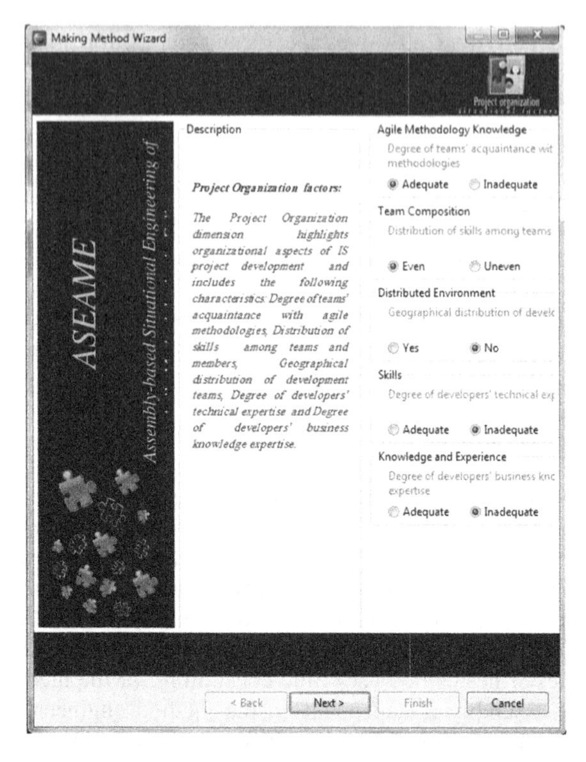

Fig. 2 Screen for specifying "project organization" situational factors in ASEAME

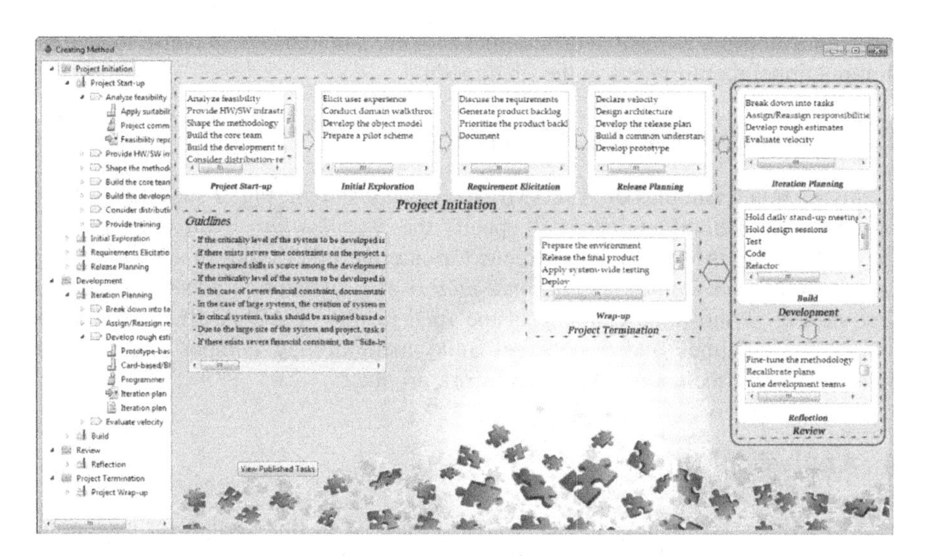

Fig. 3 Final agile method constructed by ASEAME for a specific situation

These characteristics are organized into three main groups: Functionality, Usability, and Portability. Each characteristic consists of sub-characteristics. Results of this assessment are presented in Table 2. The evaluation results indicate that ASEAME adequately provides the features that are expected in a CAME tool.

Table 2 Evaluation of ASEAME based on ISO/IEC 9126

Characterization	Sub-characterization	Support in ASEAME	Explanation
Functionality	Suitability	✓_	Does not cover all SME approaches.
	Accuracy	✓	Uses exact algorithms for selection and assembly of method fragments.
	Functionality compliance	✓	Utilizes SPEM 0.2 as a standard for method development.
Usability	Understandability	✓	Uses intelligible interfaces.
	Learnability	✓_	There is adequate documentation on the EPFC environment, but formal documentation has not been produced for ASEAME yet.
	Operability	✓	EPFC supports method development graphical notations.
	Attractiveness	✓	Graphical user interfaces are designed to enhance ASEAME's attractiveness.
Portability	Installability	✓	Based on Eclipse
	Adaptability	✓	Based on Eclipse

✓ : Adequately supported
✓_ : Weakly supported

Fig. 4 depicts ASEAME's coverage of the generic SME process, and shows that ASEAME significantly augments the EPFC tool.

We have also compared ASEAME to existing CAME tools. In comparison, ASEAME offers these advantages:

- ASEAME supports the comprehensive implementation of assembly-based SME, which is its most important contribution.
- ASEAME significantly reduces the method engineer's manual burden, as compared to other CAME tools, through enhancing automation in all the stages of SME, including requirements engineering, selection of appropriate method fragments, and method fragments assembly. Additionally, the final assessment of method fragments consistancy, (which is a very time-consuming undertaking if performed manually) is executed automatically in ASEAME.

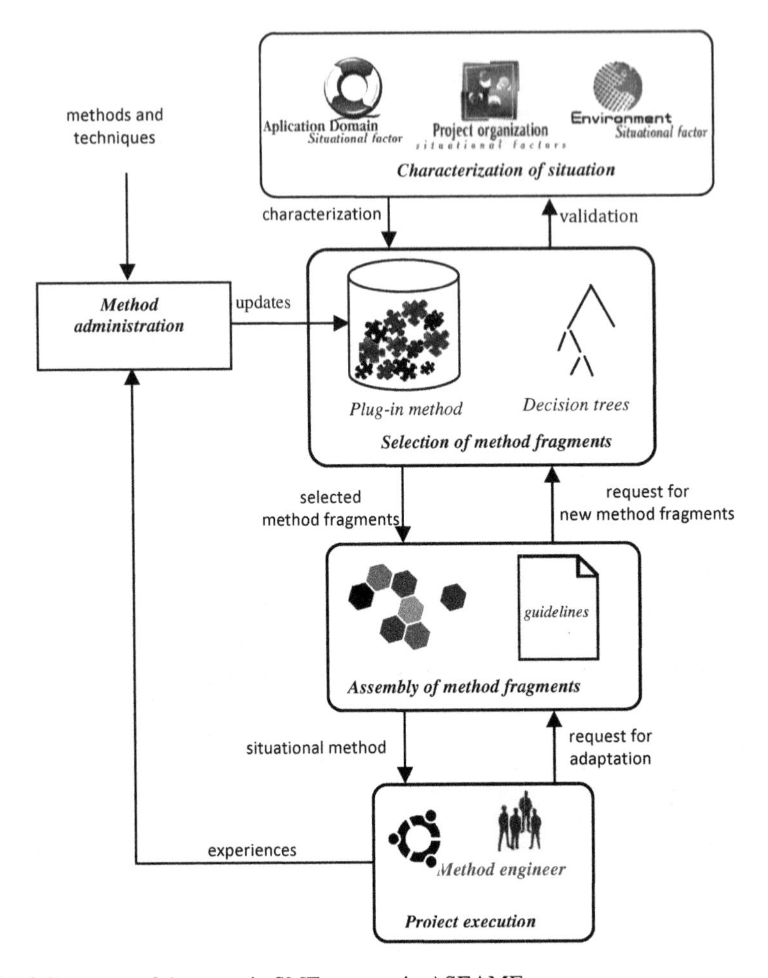

Fig. 4 Coverage of the generic SME process in ASEAME

- ASEAME provides a comprehensive set of agile method fragments derived from prominent agile methodologies, including Crystal Clear [11], DSDM [12], FDD [13], ASD [14], Scrum [15], and XP [16]. In comparison, other CAME tools lag far behind. The method fragments provided by ASEAME cover the *process* aspect as well as the *product* aspect. ASEAME supports the process aspect of methods by defining method fragments in terms of activities, tasks, and phases. The product aspect is supported by defining the related concepts in accordance with SPEM 2.0, in terms of artifacts, outcomes, and deliverables. Support for the definition of *roles* is another important feature of ASEAME.

5 Concolusions and Future Work

Our proposed CAME tool, ASEAME, enhances support for situational method engineering of agile methodologies in EPFC. Existing tools are limited in their support for comprehensive implementation of assembly-based SME, and they need a high degree of manual engagement by method engineers throughout the method construction process. ASEAME covers all the stages of SME (as defined in [1]) and reduces the method engineer's manual involvement through providing a high degree of automation. Thus, the complexities encountered in method engineering are adequately managed, production time and costs are reduced, and accuracy is enhanced.

ASEAME is based on *assembly-based* SME, and does not support other approaches, such as *paradigm-based* and *extension-based*. Therefore, the next step in this research is to extend ASEAME with support for other SME approaches. Method verification and enhancement of the agile method base are other research tasks that should be taken up.

References

1. Niknafs, A., Ramsin, R.: Computer-Aided Method Engineering: An Analysis of Existing Environments. In: Bellahsène, Z., Léonard, M. (eds.) CAiSE 2008. LNCS, vol. 5074, pp. 525–540. Springer, Heidelberg (2008)
2. Shakeri Hossein Abad, Z., Hasani Sadi, M., Ramsin, R.: Towards Tool Support for Situational Engineering of Agile Methodologies. In: Proc. Asia-Pacific Software Engineering Conference (APSEC 2010), pp. 326–335 (2010)
3. Ralyté, J., Brinkkemper, S., Henderson-Sellers, B.: Situational Method Engineering: Fundamentals and Experiences. Springer (2007)
4. Haumer, P.: Eclipse Process Framework Composer. Eclipse Foundation (2007)
5. Harmsen, A.F.: Situational Method Engineering. Moret Ernest & Young (1997)
6. Si-Said, S., Rolland, C., Grosz, G.: MENTOR: A Computer-Aided Requirements Engineering Environment. In: Constantopoulos, P., Vassiliou, Y., Mylopoulos, J. (eds.) CAiSE 1996. LNCS, vol. 1080, pp. 22–43. Springer, Heidelberg (1996)
7. Saeki, M.: CAME: The first step to automated method engineering. In: Proc. OOPSLA 2003 Workshop on Process Engineering for Object-Oriented and Component-Based Development, pp. 7–18 (2003)
8. Heym, M., Osterle, H.: A semantic data model for methodology engineering. In: Proc. Workshop on Computer-Aided Software Engineering, pp. 143–155 (1992)
9. Meta Case Consulting: Method Workbench User's Guide (2005), `http://www.metacase.com/support/40/manuals/mwb40sr2a4.pdf/`
10. Mnkandla, E., Dwolatzky, B.: Agile methodologies selection toolbox. In: Proc. International Conference on Software Engineering Advances (ICSEA 2007), pp. 72–72 (2007)
11. Cockburn, A.: Crystal Clear: A Human-Powered Methodology for Small Teams. Addison Wesley (2004)
12. Stapleton, J.: DSDM: Business Focused Development, 2nd edn. Addison Wesley (2003)

13. Palmer, S.R., Felsing, J.M.: A practical guide to feature-driven development. Prentice Hall (2002)
14. Highsmith, J.: Adaptive Software Development: A Collaborative Approach to Managing Complex Systems. Dorset House (2000)
15. Schwaber, K., Beedle, M.: Agile Software Development with Scrum. Prentice Hall (2001)
16. Beck, K., Andres, C.: Extreme programming explained: Embrace change, 2nd edn. Addison Wesley (2004)
17. Bendraou, R., Combemale, B., Cregut, X., Gervais, M.: Definition of an executable SPEM 2.0. In: Proc. Asia-Pacific Software Engineering Conference (APSEC 2007), pp. 390–397 (2007)
18. Slooten, K.V., Hodes, B.: Characterizing IS Development Projects. In: Proc. IFIP TC8, WG8.1/8.2 Working Conference on Method Engineering, pp. 29–44 (1996)
19. Henderson-Sellers, B., Ralyté, J.: Situational Method Engineering: State-of-the-Art Review. Universal Computer Science 16(3), 424–478 (2010)
20. Kornyshova, E., Deneckere, R., Salinesi, R.: Method Chunks Selection by Multicriteria Techniques: An Extension of the Assembly-based Approach. In: Ralyte, J., Brinkkemper, S., Henderson-Sellers, B. (eds.) Situational Method Engineering: Fundamentals and Experiences, pp. 64–78. Springer (2007)
21. Rolland, C., Prakash, N., Benjamen, A.: A Multi-model View of Process Modeling. Requirements Engineering 4(4), 169–187 (1999)
22. Seidita, V., Ralyté, J., Henderson-Sellers, B., Cossentino, M., Arni-Bloch, N.: A comparison of deontic matrices, maps and activity diagrams for the construction of situational methods. In: Proc. CAiSE 2007 Forum, pp. 85–88 (2007)
23. Cossentino, M., Seidita, V.: Composition of a New Process to Meet Agile Needs Using Method Engineering. In: Choren, R., Garcia, A., Lucena, C., Romanovsky, A. (eds.) SELMAS 2004. LNCS, vol. 3390, pp. 36–51. Springer, Heidelberg (2005)
24. Saeki, M.: Embedding Metrics Into Information Systems Development Methods: An Application of Method Engineering Technique. In: Eder, J., Missikoff, M. (eds.) CAiSE 2003. LNCS, vol. 2681, pp. 374–389. Springer, Heidelberg (2003)
25. Van De Weerd, I., Brinkkemper, S., Souer, J., Versendaal, J.: A situational implementation method for web-based content management system-applications: Method engineering and validation in practice. Software Process: Improvement and Practice 11(5), 521–538 (2006)
26. Brinkkemper, S., Saeki, M., Harmsen, F.: Assembly Techniques for Method Engineering. In: Pernici, B., Thanos, C. (eds.) CAiSE 1998. LNCS, vol. 1413, pp. 381–400. Springer, Heidelberg (1998)
27. International Organization for Standardization (ISO), International Electrotechnical Commission (IEC): ISO/IEC 9126: Software engineering-Product quality. ISO (2004)

Business Intelligence System in Banking Industry Case Study of Samam Bank of Iran

Maryam Marefati and Seyyed Mohsen Hashemi

Abstract. Business Intelligence (BI) is a set of tools, technologies and process in order to transform data into information and information to required knowledge for improve decision making in organization. Nowadays, we can confidently claim that the use of business intelligence solutions can increase the competitiveness of organization and outstanding it from other organization. This solution enables organization to use available information to exploit the competitive advantages of being a leader and have a better understanding of customer needs and demands to allow better communication with them. In this paper we explain about principals and elements of BI in fist section and in second section we discuss about the application of BI in banking industry and consider Saman Bank of Iran as a case study in order to applying BI solution.

Keywords: Business Intelligence, elements of business intelligence, banking industry.

1 Introduction

Business Intelligence (BI) is a set of tools, technologies and process in order to transform data into information and information to required knowledge for improve decision making in organization. Store data in data warehouse, collection and consolidation of data, reporting and data mining help us to have business intelligence. Appling of these technologies provides new ability for interact between different level of organization to reach strategic goals of it. Besides it observation

---

Maryam Marefati
Department of Computer Engineering, IAU University Arak Branch, Arak, Iran
e-mail: mmarefati@gmail.com

Seyyed Mohsen Hashemi
Dean of the Software Engineering and Artificial Intelligence Department, IAU University Science and Research Branch, Tehran, Iran
e-mail: hashemi@srbiau.ac.ir

R. Lee (Ed.): Software Engineering Research, Management and Appl. 2012, SCI 430, pp. 153–158.
springerlink.com

and analysis of key information of business process help managers to make relevant decision for corrective actions.

Nowadays we can confidently claim that the use of business intelligence solutions can increase the competitiveness of organization and distinguishing it from other organization. This solution enables organization to use available information to exploit the competitive advantages of being progressive and have a better understanding of customer needs and demands to allow better communication with them. This solution also cause that organization can control positive and negative changes and monitor them. Business intelligence is considered not only as a product or a system but also as an architecture and new approach that contains a set of analytic application which helps decision making for business process on basis of operating and analytic data base. [1]

As we consider a business intelligence solution we must first examine the principles and elements of it. The basic elements of business intelligence solution can be summed as follow:

- **Extract, Transform and Load (ETL):** If you work on business intelligence solution may be you need to store dada in data warehouse or data mart. At first it seems very simple, you must collect data from various systems and load them into data warehouse. You may force to map some columns.
  Data warehouse is a system for collection, sorting and process large volumes of data with analytic tools in order to provide complicated and meaningful information for decision makers. Data warehouse is related to ETL process. [2]
- **Data Warehouse:** "The amount of heterogeneous data that is available to organizations has made information management a seriously complicated task, yet crucial since this data can be a valuable asset for business intelligence". [3] "Nevertheless, it is believed that only about 20% information can be extracted from data warehouses concerning numeric data only, the other 80% information is hidden in non-numeric data or even in documents."[4] Data mart is a subset of data warehouse related to single business process or a single business group. Therefore, a data mart can be considered as a functional or departmental data warehouse of a smaller size and a more specific type than the overall company data warehouse. [5]
- **OLAP:** "On-line analytical processing (OLAP) systems based on a dimensional view of data have found widespread use in business applications. These systems provide good performance and ease-of-use". [6] Using multidimensional data base model lead to reduce execution time of query rather than traditional data base model (OLTP). In this model data store in special structure like cube so we can execute query quickly.
- **Data Mining:** Data mining is a tool for acquiring knowledge of the stored data. [7] Data mining, attempt to find rules, patterns and the possible desire to model, among a huge volume of data.

OLAP and data mining are complementary, for example OLAP can specify a problem in specific scope and data mining can analysis and modeling the behavior of effective elements on that scope.

- **Decision Support System:** is kind of information system that help decision maker to make decision in other word the main goal of a decision support system is "providence and help to make decision for the future".[8]

2 Application of BI in Banking Industry

Currently, huge electronic data repositories are maintained by banks and other financial institution in the world. Little information has been valuable in the repositories. Extract interested information from these huge data repositories for decision making processes by traditional and manual analysis is impossible. Business intelligence and data mining applications in the banking industry are:

- Risk Management
- Fraud Detection
- Portfolio Management
- Stock Exchange
- Customer Profiles and Customer Relationship Management (CRM)
- Anti Money Laundering

2.1 Introduction to Business Intelligence of Saman Bank

According to different needs of Saman Bank and lack of integrated management reports for bank manager at various levels, lack of applications for different offices, software team of bank began to work in different field of business intelligence that lead to design and develop different system. Each system is designed according to the needs of different departments and units.

2.1.1 Personal System

Due to the increasing activities of bank and its branches in various area of and also lack of integrated management reports in personnel system of bank, some of managers of bank and offices announce their need for a personnel system based on business intelligence. These departments are as follow:

- Bank Manager
- Personnel Office
- Financial Office
- Inspection Office
- Investigation Office
- The Formulation and Methods Office

The use case diagram of this system is:

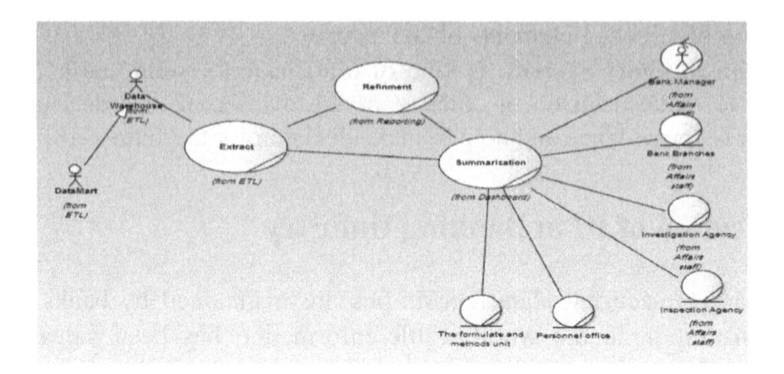

Fig. 1 Use case diagram of personnel system

2.1.2 Switch Card System

Due to the expansion of sales terminals in the country and that these services were performed by Saman Electronic Payment. Large amount of valuable information stored in electronic payment system. The switch card system is trying to extract this information to produce knowledge and acquire hidden information. The project stakeholder includes the following:

- Bank Management
- Modern Banking Department
- Monitoring Office
- Investigation Office
- Inspection Office

The use case diagram of this system display in figure 2:

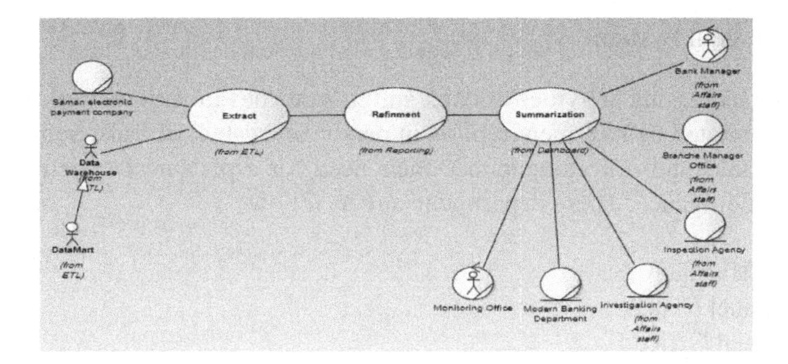

Fig. 2 Use case diagram of switch card system

In this system we must enter card number at first after confirm it, the related information of this card extract from data warehouse and filtering according to selected filter and display by different diagram. The activity diagram of switch card system show in figure 3:

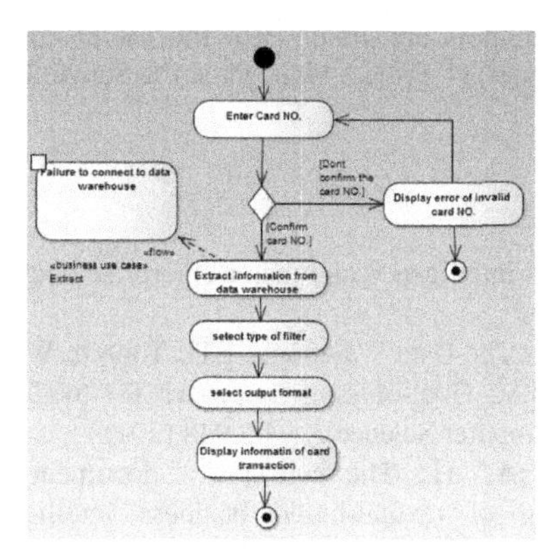

Fig. 4 Activity diagram of transaction of card in switch card system

In bellow figure we consider activities and status of ATMs.

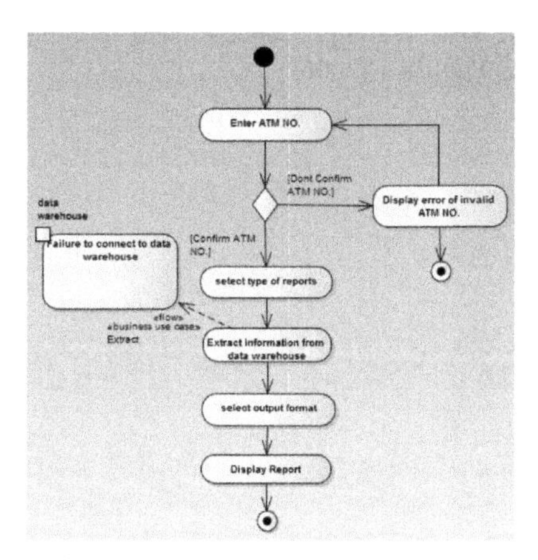

Fig. 5 Activity diagram of ATM in switch card system

3 Conclusions

The use of business intelligence systems can result in good performance on the organization's strategic goals and lead to improvement in performance of organization. In the banking industry due to intense competition between different banks provide demanded services and customer satisfaction is one of the important challenges that can be solved by using business intelligence solution. Another important application of business intelligence in banking industry is fraud detection that is to be addressed in subsequent studies to given a solution based on business intelligence.

Acknowledgments. The authors are grateful for the useful comments and explanations provided by Afshin Parhizkari, IT Project Manager at the Saman Bank of Iran.

References

1. Golestani, A.: Business intelligence and major enterprise decision. Tadbir Monthly 190 (2009)
2. Kimball, R., Caserta, J.: The Data Warehouse ETL Toolkit. Wiley Publisher (2004)
3. Ta'a, A., Abdullah, M.S.: Goal-ontology approach for modelling and designing ETL processes. Procedia Computer Science 3, 942–948 (2011)
4. Tseng, F.S.C., Chou, A.Y.H.: The concept of document warehousing for multi-dimensional modelling of textual-based business intelligence. Decision Support Systems 46, 727–744 (2006)
5. Vercellis, C.: Business Intelligence: Data Mining and Optimization for Decision. John Wiley & Sons Publishers (2009)
6. Pedersen, T., Rakic, J.G., Jenesen, C.: Object-extended OLAP querying. Data & Knowledge Engineering 68, 453–480 (2009)
7. Han, J., Kamber, M.: Data Mining: Concepts and Techniques Services, 2nd edn. Manager Simon Crump Publisher (2006)

CoDIT: Bridging the Gap between System-Level and Component-Level Development

Lukáš Hermann, Tomáš Bureš, Petr Hnětynka, and Michal Malohlava

Abstract. Component-based development traditionally recognizes two parallel views (system-level view and component-level view), which correspond to two major concerns – development of a an application and development of a reusable component for the use in application development. By having different objectives, these views have relatively disparate notion of a component, which consequently means that they are difficult (yet necessary) to combine. In this paper, we propose a method (named CoDIT), which spans the gap between the two views by providing a synchronisation between system-level view (expressed in UML 2) and component-level view. For component-level view, the method supports component frameworks with method-call as the communication style. The variability in the composition mechanisms of the component frameworks is addressed by using principles of meta-component systems. The benefits of the proposed method are evaluated on a real-life case study (in SOFA 2) together with measurements of development efforts.

1 Introduction

Component-based development (CBD) [10] has become a well-understood and widely used technique for developing of any kind of applications ranging from small single-purposed tools to large enterprise systems. CBD employs well-defined reusable pieces of code (components) and related abstractions, which are collectively embodied in a *component model*. When a component model is also

Lukáš Hermann · Tomáš Bureš · Petr Hnětynka · Michal Malohlava
Charles University, Faculty of Mathematics and Physics
Department of Distributed and Dependable Systems
Malostranske namesti 25, Prague 1, 118 00, Czech Republic

Tomáš Bureš
Academy of Sciences of the Czech Republic, Institute of Computer Science
Pod Vodarenskou vezi 2, Prague 8, 182 07, Czech Republic

R. Lee (Ed.): Software Engineering Research, Management and Appl. 2012, SCI 430, pp. 159–175.
springerlink.com

accompanied with tools and runtime environment, it forms together with them a *component framework*.

The reuse aspect, which is one of the key features of CBD, creates significant difference between non-component-based development and CBD, as CBD separates the development process of individual components (termed component-level development) from the development process of a complete application (termed system-level development) [5]. In the component-level development, the main effort is to create components to be reusable, while the system-level development focuses on looking for suitable existing components and composing them together. Thus, the system-level development is a top-down process starting at relatively high level of abstractions. On the other hand, the component-level development is bottom-up process that deals with detailed component specification.

In real-life setting these two development processes need to be run in parallel, however as they require completely different approach to components, it is not straightforward how to harmonize them [6].

In this paper we present a method that connects the system-level development process with the component-level development process. For system-level development, we propose the use of UML 2 and for the component-level development, we assume any contemporary component framework in which components communicate by calling methods on their provided/required services. Such independence of a component framework is achieved by employing principles of meta-component systems [3], which allow definition of product-lines of particular component models.

To achieve the goal the paper is structured as follows. Section 2 analyzes the individual component development and system development processes and identifies their common parts. Section 3 describes the proposed method and Section 4 evaluates it. Finally, Section 5 presents related work and Section 6 concludes.

2 Analysis of Development Processes in CBD

This section analyzes the component- and system-level developments and identifies those parts that are common for both of them. The analysis is based on [6].

2.1 System-Level Development

The system-level development is as follows. (i) First, it starts with capturing the system requirements. (ii) Then the system architecture is defined. (iii) Afterwards, the requirements of individual components are defined to allow reusing of existing components and/or developing new ones from scratch.

After the system requirements are collected in step (i), a system developer creates a system architecture based on the requirements. Simply speaking, the developer "draws" boxes (i.e. components) and links among these boxes. At this level of abstraction, it is not necessary to know exactly how services provided/required by the components are precisely defined or even implemented. The sufficient level of information is plain text that describes functionality. On the other hand, it is necessary to know which services in the architecture offer exactly the same functionality in order to be possible to merge them later and reduce a number of used components.

With the architecture defined, the developer has a complete list of necessary components and s/he can decide, which necessary components already exist and can be reused and which have to be developed from scratch. For the former group of components a mapping between their system-level view and component-level view has to exist. For the latter group of components, the developer prepares a precise specification and passes it to an individual component developer (the mapping is created after the component developer delivers an implementation of them). Finally, after all desired components are available, the whole system is composed together and can be tested and eventually deployed.

The mapping between the system-level view and the component-level view of components defines in fact how concepts on the both levels are mapped between themselves. If we assume the use of UML 2 for the system-level view, the mapping would define how the components in a particular component framework are mapped to the UML 2 components and vice-versa.

2.2 Component-Level Development

From the high level perspective, the component-level development is similar to the system-level development. (i) First, it is necessary to collect all of the component's requirements. (ii) Then the component's interfaces (i.e. the provided and required one) are defined. (iii) Finally, the component is implemented. During the implementation, the developer can reuse existing components and thus in fact "move into" the system-level development.

The interfaces of the component are defined in the form of method declarations (as stated in the introduction, we consider only component frameworks with components communicating by method calls). With the interfaces ready, the developer either implements them directly (i.e. as pieces of code in a particular programming language) or uses a set of already existing components, to which the interfaces are delegated.

2.3 Boundaries between System- and Component-Level Developments

The entities common for both the developments are the definition of the component requirements, which is typically in a plain language but more importantly the mapping between the system-level view of the component and the component-level view (in particular, mapping of components in those two views). A simplest solution would be to have a specific mapping for each component framework. However such a solution would not be practical and would require a specific set of tools for each framework. A better solution is to use concepts of meta-component systems and software product lines and to built a configurable mapping based on set of component framework features.

2.4 Example

To make the concepts more clear we show the system-level view and component-level view of a simple component framework and a corresponding mapping.

In Fig. 1, there is a system with two components connected by a interface designed in UML 2. The grey labels contain UML 2 types of the used entities. In particular, there are three component instances connected via their interfaces.

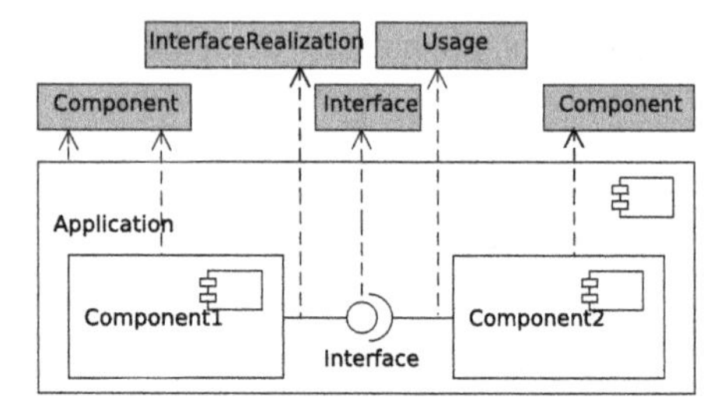

Fig. 1 UML 2 model of a system

Fig. 2 shows the implementation view of the same system in the SOFA 2 component framework [4]. Again, the labels show types the used SOFA 2 entities. In particular, there are component frames (i.e. component type that defined the set of component's interfaces), architectures (i.e. component realization) – either primitive or composite, interfaces and interface types (i.e. reusable named collection of methods on the interface). The corresponding mapping provides connection between UML 2 components and SOFA 2 component architectures and SOFA 2 component types.

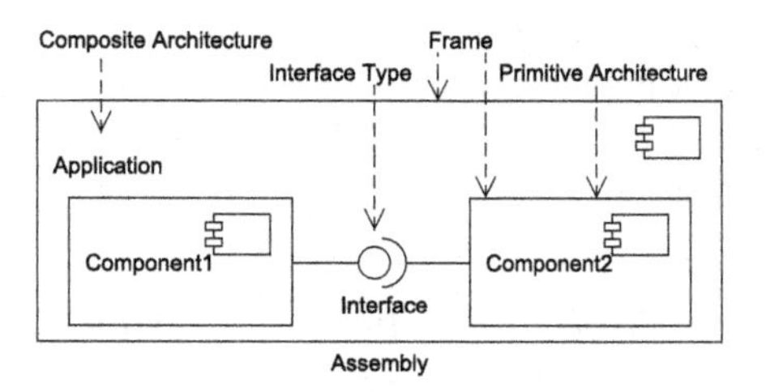

Fig. 2 SOFA 2 implementation of a sample application

3 Component Design to Implementation Transformation Method

Once we have defined both system-level and component-level views, we introduce the component design to implementation transformation (CoDIT) method in this section. The method allows spanning the gap between the two views. Essentially the method consist in mapping corresponding concepts in the two views and generating skeletons for artifacts of the component-level view. For system-level view, we rely on UML 2. For component-level view we rely on component frameworks in which components communicate by calling methods on their services (e.g. iPOJO[1], SOFA 2 [4], Fractal [2], SCA[2]). The mapping between these two views is abstracted by introduction of so called "border-line meta-model" (BDL) and "mapping meta-model" (MM). The former explicitly defines concepts of the system-level view that are relevant for interfacing with the component-level view while the latter establishes traceability between instances of border-line meta-model and entities of a particular component framework (e.g. component and component type). For brevity, we denote a particular component framework of choice as CF.

3.1 Achieving Component Framework Independence

Today, there is a plethora of component frameworks, which differ in many aspects, most importantly in ways of component composition [7]. With respect to composition, we can discern three orthogonal dimensions defining a component model: vertical composition, horizontal composition, and the number of meta-class levels for components.

In the dimension of vertical composition, two basic archetypes are found in terms of *flat* and *hierarchical* component models. A flat model allows composition only on a single level while a hierarchical one allows composition on multiple levels of nesting – see Figs. 3a and 3b. A hierarchical model therefore distinguishes two types of components – *primitive* (A, B, C in Fig. 3b) and *composite* ones (D, E in Fig. 3b) that are composed of other components (termed subcomponents).

Fig. 3 Component composition

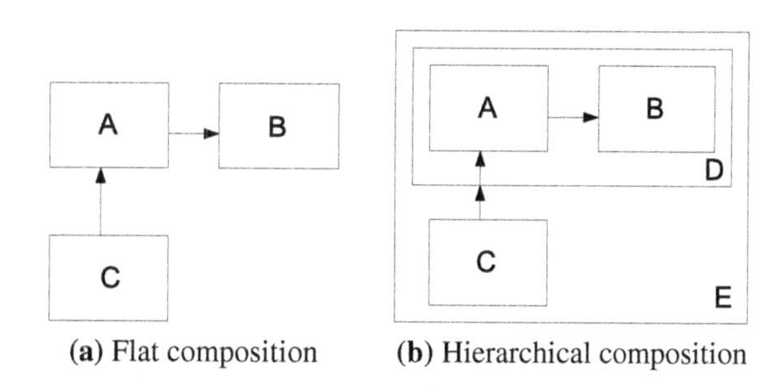

(**a**) Flat composition (**b**) Hierarchical composition

The horizontal composition dimension distinguishes how the interconnections among components are made. Two basic archetypes can be found here, namely *implicit* and *explicit* interconnections. With implicit ones, deployed components are composed automatically based on their interfaces' types and/or names (the composition can be further influenced by introduction of attributes and filters, as used by

[1] `http://felix.apache.org/site/apache-felix-ipojo.html`
[2] `http://www.osoa.org/display/Main/Service+Component+Architecture+Specifications`

OSGi[3] for instance). On the other hand, explicit interconnections have to be specified by a developer in terms of a *component architecture*.

Finally, orthogonal but related to composition is the definition of component instances. The model can essentially follow one of three archetypes: (i) singleton components only, (ii) component definitions that can be instantiated multiple times, and (iii) component definitions that can be instantiated multiple times and that is split into a definition of component type and its implementation.

To suit CoDIT method for a broad range of component frameworks in the above mentioned aspects, we use the principles of meta-component systems [3] and use the three orthogonal dimensions (along with archetypes found in each dimension) as features collectively defining a component model. Using such parameterization, we derive component models by different combinations of archetypes.

In our approach, we use three-letter structured acronyms for such component models. First letter signifies vertical composition: **F**lat, **H**ierarchical. Second letter signifies horizontal composition: **I**mplicit, **E**xplicit. Third letter is a number of meta-class levels for components (i.e. **0** – components as singletons, **1** – components and their instances, **2** – component types, components, and component instances).

Although the CoDIT method may be adjusted for any combination of archetypes, we describe here (due to space constrains) CoDIT only for those combinations of archetypes for which a well-known corresponding component framework exists. In particular, we consider the following combinations: FI0 (e.g. OSGi), HI1 (e.g. iPOJO), HE1 (e.g. Fractal), and HE2 (e.g. SOFA 2).

Formally, CoDIT comprises of two steps: (i) from UML 2 model to BDL, and (ii) from BDL to CF. The first step allows us to move from UML 2, which is unnecessarily rich, vague and with only implicitly given semantics to a model (BDL), which is simple and contains no unnecessary concepts and the semantics of which is explicitly defined (Sect. 3.2). The second step comprises creation of component framework (CF) artifacts. Since the second step is CF-dependent, we describe its variations separately for each of the considered archetypes of component models (i.e. FI0, HI1, HE1, and HE2).

3.2 BDL Meta-Model and Its Relation to UML 2

The *border-line meta-model* (BDL) allows capturing a hierarchical composition of components. The meta-model of BDL is shown in Figure 4. Essentially a component, which is an arbitrary large encapsulated piece of software is captured as an instance of the class `Component`. Each component may interact with the outside world only via interfaces (captured as instances of the class `Interface`). An interface is a named instance of a collection of methods. Depending on whether the method is offered or called by the component, the interface is said to be *provided* or *required* respectively. We further demand that interface names are unique per component. In a similar way, we require component names to be unique per BDL.

[3] `http://www.osgi.org`

For components, we allow for components composed of other components. The composition is realized via the association `subcomponents` of the class `Component`. Each subcomponent communicates only via its interfaces. The routing of the calls on interfaces is specified in terms of instances of the `Connection` class. Each connection constitutes a communication path from a provided interface to a required interface transmitting all calls between the interfaces.

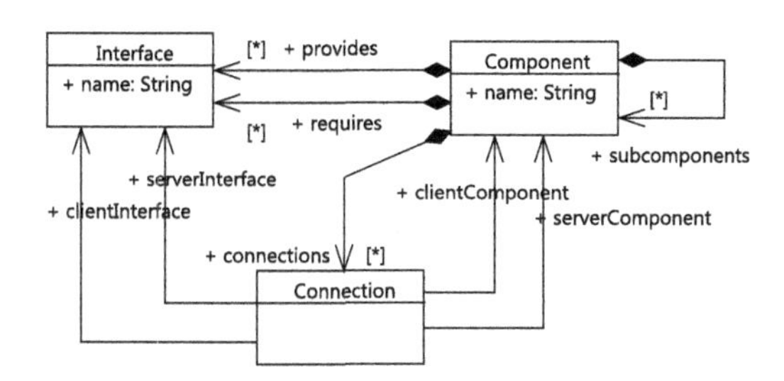

Fig. 4 BDL meta-model

Having introduced the BDL meta-model, we overview a set of concepts used by the CoDIT approach to denote the system-level view in UML 2. Basically, we use a subset of a UML 2 component diagram (see example in Figure 1).

Formally speaking, the UML 2 class `Component` corresponds one-to-one to the BDL class `Component`, with the attribute `name` having identical role in both the meta-models and with the composition `ownedElement` in UML 2 corresponding to the composition `subcomponents` in BDL. Regarding interfaces, UML 2 distinguishes `Interface` (graphically represented as a circle) and `InterfaceRealization` (graphically represented as a line connecting the circle and the component, with an open arc for required interfaces). The mapping to BDL transforms the UML 2 `Interface` together with attached `Interface-Realization` to an instance of the BDL `Interface` referenced by the association `provides`. Similarly it transforms the UML 2 `Usage` to an instance of the BDL `Interface` referenced by the association `requires`. The attachment of `Usage` to `Interface` is transformed to an instance of the BDL `Connection` with the association `client` pointing to the BDL requires interface and the association `server` pointing to the BDL provides interface.

3.3 *Transforming BDL to CF*

Once the BDL representation is obtained from the UML 2 model, it is transformed to artifacts of CF. Since this process needs additional information not present in BDL (e.g. signatures of interfaces, component types), we introduce the *mapping model* (MM), which adds the missing information. By its nature, the mapping model is CF-dependent, thus we deal with it separately based on a particular selection of component model archetypes.

CoDIT method infers MM when invoked for the first time, then it allows for adjustment of MM by a developer. This makes it possible for instance to reuse existing CF artifacts or specifying that a few component instances in BDL stem from one component implementation. After the adjustment, CF artifacts are automatically generated (according to rules described in the rest of the section). These steps may be reiterated any number of times necessary for development of the resulting system. A schematics of the approach is depicted in Figure 5.

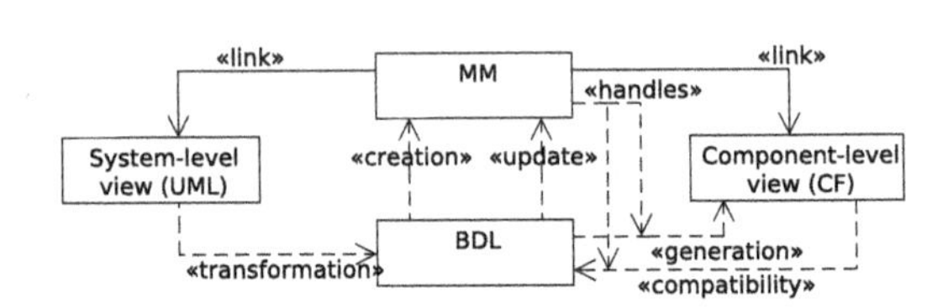

Fig. 5 Relations between models

1. When no mapping exists in MM, default mapping is inferred, stored in MM and a corresponding CF artifact is generated.
2. When mapping exists in MM but the mapping and/or mapped CF artifacts are not up-to-date (in terms of its interfaces, their types, subcomponents and/or connections among them) with respect to BDL, MM and corresponding CF artifacts are updated.
3. When mapping exists in MM and it is up-to-date including corresponding CF artifacts, nothing happens.

To describe the transformation of BDL to CF in detail, we define in this section an abstract core of a component model for a given combination of component model archetypes. Although such a core of a component model (CoCM) is different to the metamodel of a real component framework (CF), it can be easily shown that the mapping between CoCM and the metamodel of a corresponding real CF is essentially just renaming classes and attributes. (This is because of the fact that the component model archetypes define major portion of the composition features of the component model.)

3.4 FI0 Family

We start with description of FI0 family, which corresponds to flat component models with implicit connections and no distinction of component instances. The most well know representative of this family is OSGi. CoCM of this family is depicted in Fig. 6 while the corresponding MM is shown in Fig. 7.

Synchronizing MM from BDL is straightforward as in the case of FI0 family, MM features basically no additional information. It only provides one-to-one mapping between BDL components and CoCM components (based on unique

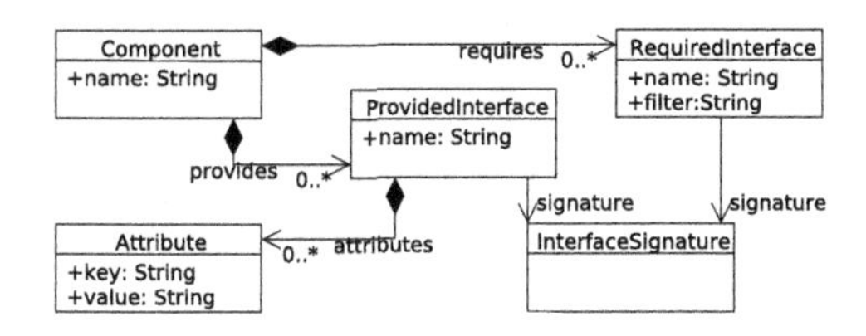

Fig. 6 FI0 CoCM

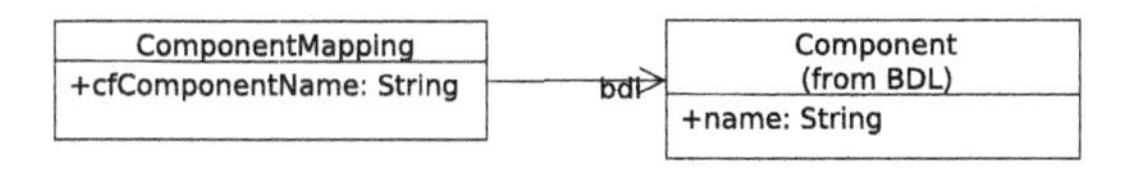

Fig. 7 FI0 MM

component names[4]). The key role of the mapping is to allow reuse of existing components. By mapping components, the mapping also implicitly relates together interfaces via equality on interface names.

Synchronizing CoCM with BDL is also relatively straightforward – consisting mainly in generating CoCM components according MM, which for non-customized MM means a CoCM component for every BDL component. The only difficulty comes from the fact that BDL relies on explicit connections (as these are natural in system-level design) while CoCM for FI0 family relies on implicit connections. This is solved by employing filters and attributes. We associate each provided interface in CoCM with an attribute *bdl-id* which has value equal to a string formed by pattern *<component-name>–<interface name>*. The connected required interface is equipped with a filter that filters out other provided interfaces except the one with the corresponding *bdl-id* attribute. When sticking to the LDAP syntax (employed e.g. in iPOJO), the filter would be of type "`(bdl-id=...)`".

3.5 *HI1 Family*

HI1 family is basically an extension of FI0 family, which features hierarchical components and multiple instances of a single component. The best known representative of this family is iPOJO, which is an extension by Apache to OSGi services.

CoCM of HI1 family is depicted in Fig. 8. Compared to CoCM of FI0 family, there is a class for representing component instances. To specialize particular instances, each instance may be configured with a number of attributes (essentially key-value pairs). The possibility of hierarchical components is realized by the association `subcomponents`.

[4] Component names (as opposed to model links to CoCM) are used in order to support adjustment of MM before CoCM and related CF artifacts are generated.

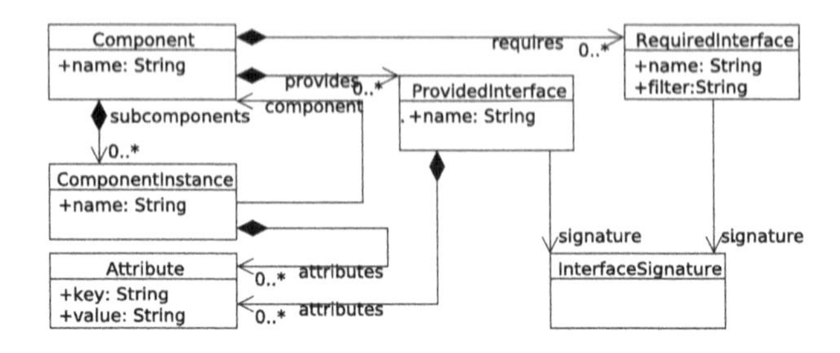

Fig. 8 HI1 CoCM

MM of HI1 family remains the same as in the case of FI0 family. Mapping of component instances is implicit, based on names of component instances.

Synchronization of CoCM and MM with BDL works as an extension of the synchronization of FI0 family (i.e. all the steps of synchronization that hold for FI0 family are preserved). In addition to FI0, there is an instance of the class `ComponentInstance` created for each instance of the class `Component` in BDL. Additionally, a subcomponent relationship is established in the same way as captured by BDL.

3.6 HE1 Family

HE1 family corresponds to the most typical hierarchical component models with explicit architecture (e.g. Fractal and its derivatives). These models allow several instantiations of a component and for hierarchical nesting of components (similarly to HI1 family). In contrast to HI1 family, the connections are explicit, which means that each connection between component interfaces has to be specified in the component architecture (typically the connections are specified as interconnects of sub-components and delegations to super-component's interfaces).

CoCM corresponding to HE1 family is shown in Fig. 9. The semantics of the classes is the same as in the case of HI1 family. Only the classes `Provided-Interface` and `RequiredInterface` lack association to Attribute and the filter, respectively. This is substituted by class Connection, which defines a communication link between a particular interface on a client component instance (i.e. the one with requires role) and a particular interface on a server component instance (i.e. the one with provides role). HE1 CoCM constraints the connections only to be between required and provided interface of direct subcomponents of the same super-component, or to be between required-required or provided-provided pair of interfaces of a subcomponent and the super-component (this corresponds to a delegation). MM of HE2 family remains the same as in the case of FI0 family. Again, mapping of component instances is implicit, based on names of component instances.

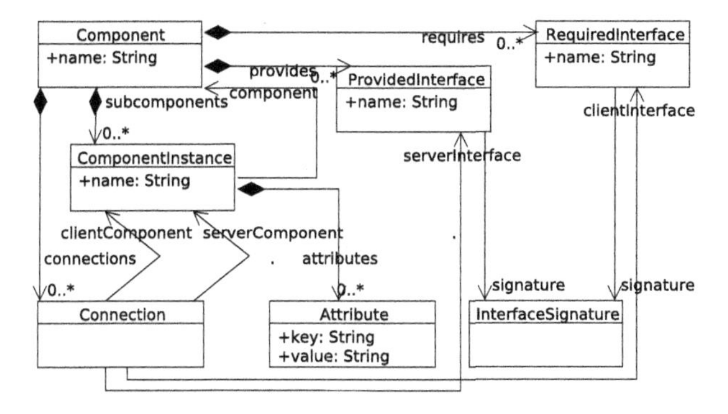

Fig. 9 HE1 CoCM

Synchronization of CoCM and MM with BDL works in the same way as for HI1 family, only connections in BDL are not mapped to the *bdl-id* attribute and filter, rather they are directly (in fact 1:1) mapped to instances of CoCM class `Connection`.

3.7 *HE2 Family*

HE2 family has the same characteristics as HE1 family, only it adds an explicit notion of reusable `ComponentType`. As such, HE2 family allows product-lines by allowing subcomponents to be specified also by stating solely the component type. (In HI1 family, subcomponents have to be specified only pointing to a component.) Consequently, a subcomponent may be specified as a black-box without elaborating on internal functionality. The particular implementation of a subcomponent is specified when the whole system is being composed and its architecture is being refined. The most distinguished representative of HE2 family is SOFA2 and its derivatives (e.g. SOFA-HI).

CoCM of HE2 family is shown in Fig 10. The semantics of the classes is the same as in the case of HE1 family. The distinction (as described in the paragraph above) lies in the existence of additional class `ComponentType`. This class carries the black-box specification of a component (including the interfaces, which are consequently not associated from the class `Component`). The support for black-box (i.e. component type-based) specification of a subcomponent is reflected by the `componentType` association from the class `ComponentInstance` to the class `ComponentType`. There is additional "xor" constraint between the `componentType` and `component` associations expressing that a component instance has either assigned an implementation (in terms of a specific instance of the class `Component`) or has only its interfaces specified without concretizing its implementation (i.e. has assigned a component type).

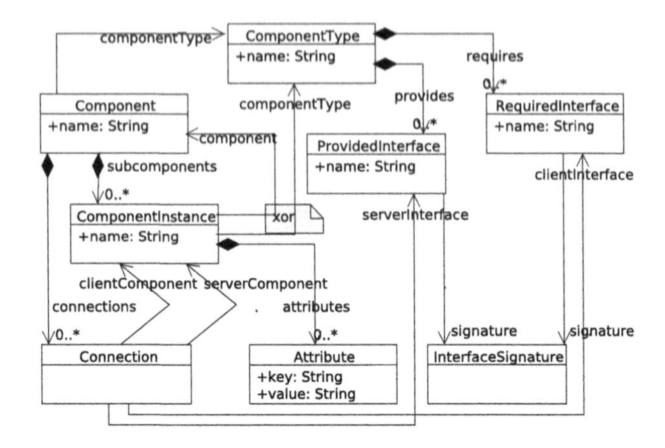

Fig. 10 HE2 CoCM

Compared to HE1 family, MM for HE2 (see Fig. 11) explicitly associates a BDL `Component` with CoCM `ComponentType` using a unique name of the component type in CoCM. The reference to a specific CoCM `Component` is changed optional (i.e. the attribute `cfComponentName` may be null). This allows for: (i) specifying a sub-component as a black-box (i.e. with `cfComponentName` attribute is null), and (ii) influencing a component type for CF component to be generated (i.e. with `cfComponentName` attribute as not null).

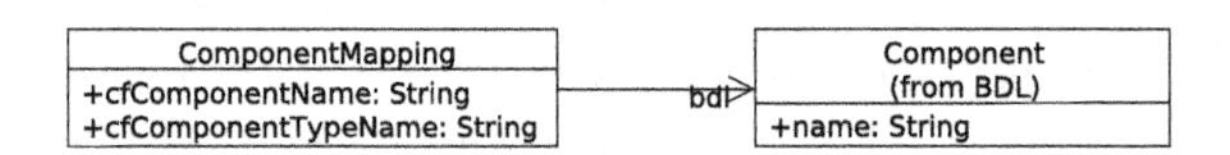

Fig. 11 HE2 MM

Mapping of HE2 CoCM and HE2 MM with BDL is based on the synchronization in HE1 family. Only a CoCM component type is created for each component type name in MM. Further, when a BDL subcomponent has a mapping in MM that does not state the component name, it is translated to CoCM as a subcomponent specified by a component type.

4 Evaluation and a Case-Study

To evaluate the CoDIT method, we have implemented it for the SOFA 2 component framework including support for Eclipse IDE. Additionally, we tested the implementation for SOFA2 on a real-life use-case and evaluated the development effort.

The SOFA 2 component framework has been selected for evaluation because it falls into HE2 family, which can be argued as the most complex family among those with explicit interconnections. This can be easily shown by realizing that nested models form a superset of flat models and that 2-level meta-class hierarchy used in HE2 subsumes lower-level classes (i.e. 0 and 1).

In implementing CoDIT for SOFA 2, we realized both the transformation steps (i.e. from UML 2 to BDL, and from BDL to SOFA 2 CF) via QVT transformations [9]. These were complemented by EMF[5] meta-models and OCL validation constraints based on those presented in Section 3. In case of the second step, we skipped creating CoCM for HE2 family and implemented the QVT transformation directly to SOFA 2 meta-model. This was straightforward as corresponding parts of SOFA 2 meta-model (see [4]) are structurally almost identical to CoCM of HE2 family.

To strengthen the benefits of CoDIT method on SOFA 2 development, we have implemented support for CoDIT as a part of SOFA 2 IDE, which is an Eclipse-based tool for development of SOFA 2 applications.

The use of CoDIT in SOFA 2 IDE consists of the following steps: (i) the developer creates a UML 2 model with the Eclipe UML 2 editor (this tool is part of the Eclipse modeling project); (ii) the developer chooses to create a new SOFA 2 Project from the UML 2 model; (iii) CoDIT validates the UML 2 model and if the model is valid, it creates a mapping model, (iv) the mapping model editor (see Figure 12) opens it and allows the developer to specify mapping for components and interfaces; (v) with the mapping complete, the CoDIT method generates SOFA 2 implementation artifacts (i.e. component architectures, component types, interface types, etc.); (vi) as mapping is stored in a file, the developer is able to open it at any point and synchronize it with the UML 2 model and/or update the generated SOFA 2 implementation artifacts.

CoDIT support for SOFA 2 is publicly available as part of the SOFA 2 component framework[6].

To evaluate the impact of the CoDIT method on development of a component system, we have created a real-life use-case (Figure 13). It models a simplified information system of a public library. The library component has 3 subcomponents – the library system and user and administration terminals. The library system is further composed of other 2 components – the user registry and store. The user registry manages a list of users and handles user requests. If a request is authorized it is stored in the store component. The store component itself manages also information about books.

In a real implementation of this example, the hierarchy would be deeper, however to evaluate our method, the two levels are enough.

We developed the library example by hand and using the CoDIT method and compared the required development effort. We measured the overall time needed for development, number of user interactions with IDE (mouse clicks, drag&drops,

[5] `http://www.eclipse.org/EMF/`

[6] `http://sofa.ow2.org`

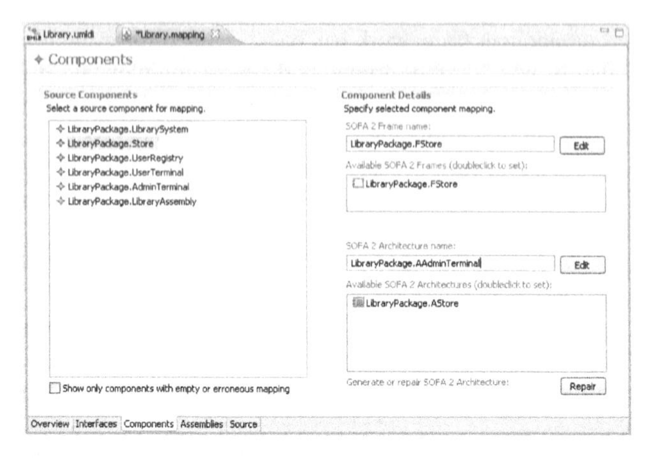

Fig. 12 Component page of the mapping model editor

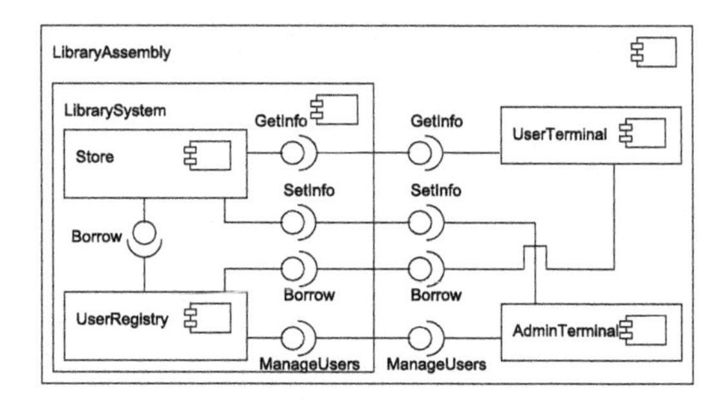

Fig. 13 Architecture of example application

etc.), and the “system work time” (i.e. overall time of a user is waiting for the system response).

The results are shown in Table 1 in the column “1st”.

Second experiment (the column “2nd” in the table) performed measurement of a change of the existing system (particularly, the library system architecture was updated by a new interface used by the user terminal component).

Third experiment (the column “3rd” in the table) was a development of the whole library example as in the first experiment but this time all components but the store and user terminal were already developed and reused.

As can be seen from the results, the best case for the CoDIT method is development from scratch (first experimental) when it saved significant amount of time and user activity and even overall system work time (as most of the code was generated in a single batch). The CoDIT method is also good in case some components are already available for reuse (third experiment). For changing an existing system (second experiment), the results of the CoDIT method are the worst but still in most cases slightly better than manual development.

Table 1 Results of measurement

	1st	2nd	3nd
Overall time (s)			
With the CDIT	55	100	128
Without the CDIT	758	153	603
With/Without (in %)	7	65	21
User interactions (count)			
With the CDIT	11	28	34
Without the CDIT	326	58	261
With/Without (in %)	3	48	13
System work time (s)			
With the CDIT	35	38	23
Without the CDIT	56	10	39
With/Without (in %)	63	380	59

5 Related Work

There is in fact no such other method that is independent on particular component-level framework. There are many works allowing generation of implementation in e.g. EJB, CCM from UML 2 components, but all of them are specific for the implementation framework, which cannot be simply changed. Typically, these works are directly parts of UML 2 editors.

The paper [8] introduces a mapping from UML 2 to the old version of the SOFA framework and Fractal component model [2]. However the system-level development is closely tied with the component-level development and thus the method is not component framework independent.

There is also a paper [1] that analyzes mapping between the UML state machine diagrams and behaviour protocols for the Fractal component model and that introduces mapping between the UML component diagrams and Fractal ADL. However there is no support of reusability and propagation of changes from the system view to the component view.

In the case we do not limit ourselves to component-based development, we can find projects with similarities to our method.

All model-driven development tools are related to our method as they allow transformations between different levels of view on a developed system. But again, there is necessity to prepare specific transformation for each combination of input and output levels.

Another related projects are for example reflective middlewares like *Dream*[7] that provide highly configurable middleware via the same means like meta-component systems we employ. Similar approaches exist also in the area of enterprise systems;

[7] `http://dream.objectweb.org/`

for example the *Spring*[8] project, which offer a very configurable run-time with extension abilities.

6 Conclusion

In this paper, we have presented the CoDIT method for spanning the gap between the system-level view and the component-level view. In particular we rely on UML 2 for the system-level view and method-call based component frameworks for the component-level view. The component framework independence of the method is realized by applying principles of meta-component systems and by designing three dimensional space of compositional archetypes. Based on this, we have defined representative abstract component models that represent different families of component systems. Subsequently, we have described specialization of our method for each of the relevant families. Our method has been evaluated on a real-life case-study for a representative of the most complex family of component systems with explicit connections. The results of the measurements of the development efforts show that our method saves at least 90% of effort connected with creation of component in the initial phase and at least 35% of effort in modifications. Also, our method supports reuse of components in which case it is still able to provide benefits in the order of 10% of saved development efforts.

Acknowledgements. This work was partially supported by the Grant Agency of the Czech Republic project P103/11/1489, by the Charles University grant SVV-2012-265312, and by the Czech Science Foundation grant 201/09/H057.

References

1. Ahumada, S., Apvrille, L., Barros, T., Cansado, A., Madelaine, E., Salageanu, E.: Specifying fractal and GCM components with UML. In: Proceedings of SCCC 2007, Iquique, Chile, pp. 53–62. IEEE CS (2007), doi:10.1109/SCCC.2007.21
2. Bruneton, E., Coupaye, T., Leclercq, M., Quema, V., Stefani, J.: The FRACTAL component model and its support in Java. Software: Practice and Experience 36(11-12), 1257–1284 (2006), doi:10.1002/spe.767
3. Bures, T., Hnetynka, P., Malohlava, M.: Using a product line for creating component systems. In: Proceedings of the 2009 ACM Symposium of Applied Computing (SAC 2009), Honolulu, Hawaii, USA pp. 501–508. ACM Press (2009), doi:10.1145/1529282.1529388
4. Bures, T., Hnetynka, P., Plasil, F.: SOFA 2.0: Balancing advanced features in a hierarchical component model. In: Proceedings of SERA 2006, Seattle, USA, pp. 40–48. IEEE CS (2006), doi:10.1109/SERA.2006.62
5. Crnkovic, I., Chaudron, M., Larsson, S.: Component-based development process and component lifecycle. Journal of Computing and Information Technology 13(4), 321–327 (2005), doi:10.2498/cit.2005.04.10

[8] http://www.springframework.org/

6. Crnkovic, I., Larsson, M.: Building reliable component-based software systems. Artech House, Norwood (2002)
7. Lau, K., Wang, Z.: Software component models. IEEE Transactions on Software Engineering 33(10), 709–724 (2007), doi:10.1109/TSE.2007.70726
8. Mencl, V., Polak, M.: UML 2.0 components and fractal: An analysis. In: 5th International ECOOP Workshop on Fractal Component Model, Nantes, France (2006)
9. OMG: Query/View/Transformation, v1.1. OMG document formal/2011-01-01 (2011)
10. Szyperski, C.: Component software: beyond object-oriented programming, 2nd edn. Addison-Wesley, Boston (2002)

A Resource Scheduling Algorithm Based on Trust Degree in Cloud Computing

Mingshan Xie, Mengxing Huang, and Bing Wan

Abstract. Cloud resource architecture is put forward based on servers according to the characteristics of Cloud Computing. And then the trust degree for the resource scheduling in Cloud Computing is defined and a resource scheduling algorithm based on trust degree is presented. Finally, the functional characteristics of this scheduling algorithm are analyzed by simulation, and the simulation results show that the resource scheduling algorithm based on trust degree in Cloud computing possesses the better stability and low risk on completing tasks.

Keywords: Cloud computing, resource scheduling, trust degree.

1 Introduction

Cloud Computing is an entirely new computing service mode which is rose in the 21st century. With the development of Internet services providers only need to deploy application resources on the server, customers will be able to order services they need via the Internet from application services providers, and pay for them according to the number of services and using time. This not only saves the cost of customers in the purchase of products, technical training and something else, but also greatly reduces the threshold and risk of informatization for small and medium enterprises. Since with the growth of profession customers' requirements expand, leading service providers need to deploy different applications on the server and manage them, using this platform to help service providers gain huge profits. However, therefore service providers also have to set up multiple servers in parallel in different areas. On the fact of that, as for the rational utilization of network resources and to provide reliable services will become increasingly concerned by service providers and become a key issue. Cloud Computing has been popularized

Mingshan Xie · Mengxing Huang · Bing Wan
College of Information Science & Technology
Hainan University
Haikou, P. R. China
e-mail: huangmx09@gmail.com

R. Lee (Ed.): Software Engineering Research, Management and Appl. 2012, SCI 430, pp. 177–184.
springerlink.com

in 2009, after the raising of Google API services, numerous software vendors have launched their own Cloud Computing projects: Microsoft firstly proposed the "Azure" program, IBM also established the first Cloud Computing Center in China, AMD, Inspur, Cisco and other vendors are gradually starting to develop Cloud Computing platform as a commercial operation mode.

In recent years lots of researches on scheduling of resources were put forward from different perspectives, of which Round Robin (RR), First Come First Service (FCFS), and Fixed-Priority Scheduling Strategies and so on are more traditional approaches. Song yuan Jun etc. [1] taking into account both the execution time of tasks and the value of tasks proposed one cloud scheduling algorithms for multi-machine and multi-tasking real-time system. This algorithm effectively solved the issue of determining the priority of task and timing issue, which improved system resource utilization efficiency, quality of operation and effectiveness of the whole system. Sun Ruifeng etc. [2] based on the virtualization of Cloud Computing, presents a resource scheduling strategy combining lease and dynamic multi-level resource pool, which is able to effectively reduce the idle time of resources and improve the utilization rate of resources. LIU Zhi-jia etc. [3] proposed an user-expected task function for Cloud Computing resource scheduling, and according to user-expected function and the task classification of service quality proposed scheduling algorithm based on user-expected function. Rodrigo N. etc. [4] describes the architecture and features of Cloud SIM that is one simulation platform for Cloud Computing. A. Benoit etc. [5] proposed the tree structure of users and resources.

Among of these literatures, the literature [5] only takes the allocation of resources as a starting point, and doesn't make a choice of the stability of scheduling resources. This paper put forward Cloud resources architecture based on the insufficient of reference 6 and the situation of services provided by Cloud Computing. And then the paper defined the Trust degree for the resource scheduling in Cloud Computing on the basis of these factors, and according to the size of trust degree completed tasks scheduling. This paper mainly do some researches as follows: (1) managing and scheduling all resources unified in Cloud; (2) defining the trust degree for the resource scheduling in Cloud Computing and sorting the resources for scheduling according to the trust degree; (3) by means of analyzing this scheduling algorithm and simulation proved that the scheduling strategy based on trust degree under Cloud environment has better stability and can get low risk on completing tasks.

2 Designing the Cloud Computing Resources Pool

2.1 Resources in Cloud Computing

Computing resources are nothing more than all five things that, CPU, RAM, Disk, RAM-Disk IO and Network, etc. And resource scheduling is simply optimizing the utilization of these five resources to complete the allocated work in the shortest possible time.

Compared to Grid Computing and Distributed Computing, Cloud Computing has obvious features. Firstly, its low cost is the most outstanding features. Secondly the virtual machine support makes some difficult things in network environment easier to handle now. Thirdly the execution of mirror deployment can make the implementation of the heterogeneous procedures' interoperability which is difficult to deal with in the past have become easier to handle. Fourthly the emphasis on service-oriented has some new mechanisms particularly the mechanism of more suitable for commercial operation.

Cloud and Grid both are designed to be able to very scale efficiently horizontally. Both of them are able to withstand the failure of an individual element or node and are charged by usage. However, Grid is usually processing batch jobs and has a clear start and end while Cloud Service can run continuously. In addition, Cloud enlarges the type of available resources including file storage, database and Web services, and will extend to the Web and business applications.

2.2 Designing the Structure of Resource Pool Based on Virtualization in Cloud Computing

According to the virtualization of Cloud Computing, network resources such as servers and storage devices are integrated and then segmented to allocate resource on demand and increase automatically. As for the controlling end, Cloud Computing regards all IT resources as a Resource Pool. But it distributes resources to different resource pools according to the physical attributes, which means that virtual machines not only possess reliable processing power but also obligate some space for storage and Internet access. Utilizing the virtualization of Cloud Computing, resources are virtualized into m slots, according to a certain common property of resources (such as CPU type, operating system and memory size, etc.) all resources will be classified to form a resource pool, as it shown in Figure 1.

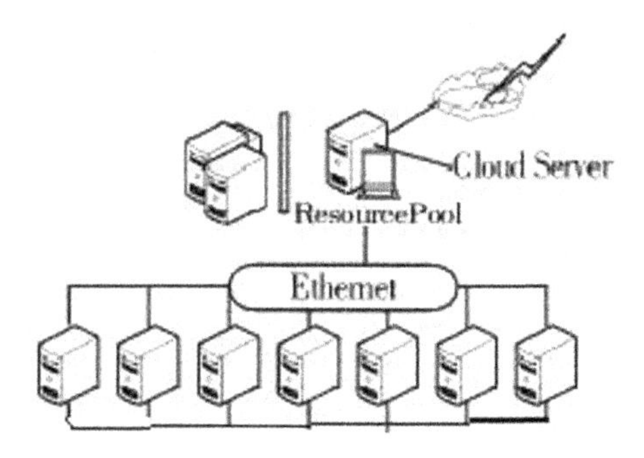

Fig. 1 The structure of resource pool

Considering network or other factors may cause servers to crash or other failures, so it uses hot standby technology to realize file backup and configure several candidate servers.

2.3 Defining Trust Degree of Computing Slots

There are lots of computing slots in Cloud Computing, and their physical locations are distributive. Because the conditions diverse from each other in different places, it is difficult to ensure that each slots can run correctly, but fortunately Cloud Computing adjusts dynamically. This paper analyzes the demand for the number of resources in Cloud Computing by proposing the Trust degree for the allocation of resources. For the user tasks that requiring high reliability, choosing resources of high reliability to serve for them can dramatically improve the service quality.

Assumed the Trust degree of virtual computing slots is X_{trust}, the number of failures since the virtual slot has been put into resource pool is C, and the number of tasks that has been executed since the virtual was put into resource pool is W, then the probability of a computing slot's failure is P = C / W, which is a real number less than 1. To facilitate sorting, it translates P into a positive number, so Trust degree can be defined as:

$$X_{trust} = -\ln(C / W)$$

3 A Scheduling Model Based on Trust Degree

3.1 A Resource Scheduling Model in Cloud Computing Based on Trust Degree

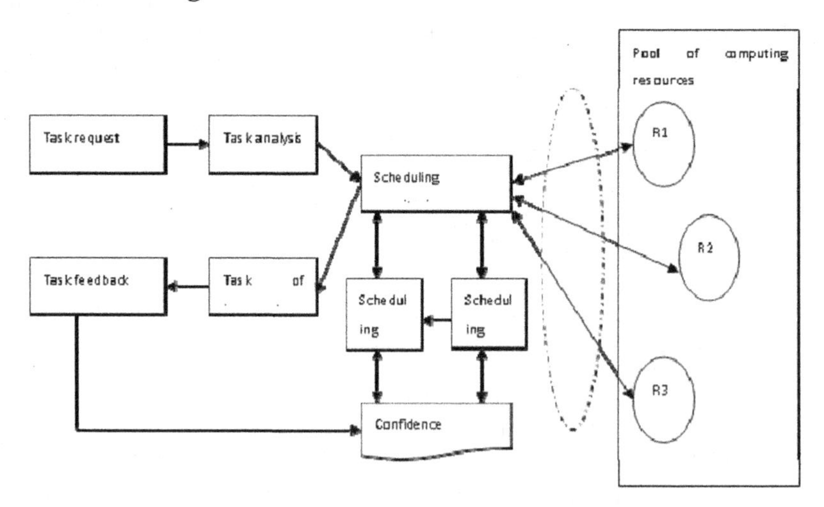

Fig. 2 A resource scheduling model in Cloud Computing based on trust degree

In which R1, R2, R3, ... represent the virtual computing slots; scheduling adjustment and scheduling logs communicate with scheduling monitoring at any time, and scheduling logs registered in scheduling in a table that stores trust degree at any time; sorting trust degree descendingly and calling computing slots in order that is calling the slot whose trust degree is greater firstly.

3.2 The Process of Resource Scheduling

Step 1: Tasks request access to servers; analyzing the tasks; the scheduling monitoring mechanism distributes every task computing slot by trust degree descendingly.

Step 2: Scheduling monitoring mechanism monitor every computing slots' operation timely, it calls scheduling adjustment and take leasing measures to allocate computing slots for every task by trust degree descendingly in spare slots and compute complementally.

Step 3: Scheduling monitoring mechanism computes the trust degree of the failure slots and stores them into the trust degree record table. And for the leasing spare slots, it computes the trust degree and stores them into the trust degree record table too.

Step 4: All slots return the results to servers and integrate all tasks; the task feedback adds the number of completing tasks into the trust degree record table.

4 Simulation and Analysis of the Algorithm's Functionality

There are four different kinds of factors influencing the completion of tasks; they are completion time, bandwidth, reliability and costs. The completion time and reliability mainly determine the trust degree directly. The completion time for an individual task is:

$$T_F = T_{wait} + T_{exec} + T_{trans}$$

T_{wait} is the waiting time from being submitted to being allocated resources of task. T_{exec} is the execution time of one task. T_{trans} is the transmission time of one task. Assumed that every slot in resource pools has a same resource copy and same processing capacity, that is so say that every server of the model can achieve the same results from any computing lots. There are so many tasks in Cloud Computing, then if assumed that there're n tasks, the total time for computing these n tasks is nT_F. If there is something wrong with the computing slots, it will lease the spare slots so that it will increase the task's waiting time T_{wait}. Additional computing will increase the execution of task T_{exec}. Accordingly, the task needs to be called again which will increase the transmission time T_{trans}. So obviously TF will increase accordingly.

Then the total time of executing k tasks again for each slot is:

$$T^{'} = k\mathrm{T}_{\mathrm{F}} + \Delta T$$

ΔT is the increment of T_F when a single computing slot goes wrong, and k is the amount of executing tasks.

Assumed the load capacity of each slot is same and all are the upper limit N. That is no more tasks will be accepted while the number of the tasks is greater than N, and then server will allocate the remaining tasks to other computing slots.

To make experimental comparison obviously simple, this paper takes three computing slots as is shown in Figure 3. A is a server, so the server has three computing slots to choose. This paper use Matlab to model and set up the simulation platform. In this model, the total number of executing tasks at one certain moment is n. Assumed that the increment of T_F when a single computing slot goes wrong is double T_F. Then the risk time of a single computing slot executing k tasks is: $T^{''} = k\mathrm{T}_{\mathrm{F}} + P\mathrm{T}_{\mathrm{F}} = (1+P)\mathrm{T}_{\mathrm{F}}$

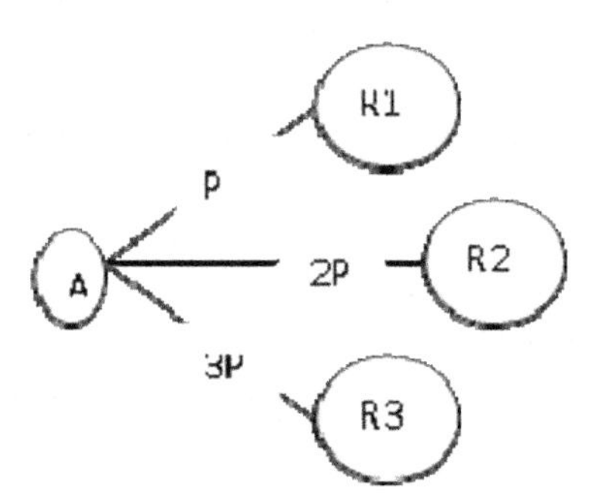

Fig. 3

Assumed that the error probabilities of R1, R2 and R3 are P, 2P, and 3P respectively. The Trust degree of R1, R2 and R3 is Xtrust(R1)> Xtrust(R2)> Xtrust(R3).

In the case of not sorting the Trust degree of computing slots, since each slot has the opportunity to provide resources to server, so the risk time of completing tasks is T_1:

$$T_1 = n((1+P)+(1+2P)+(1+3P))T_F/3 = n(1+2P)T_F/3 \quad n \in anyvalue$$

Sorting the Trust degree of computing slots, in worst case the risk time of completing tasks is T_2:

$$T_2 = \begin{cases} n(1+P)T_F & while..n \le N \\ n((1+P)+(1+2P))T_F/2 = n(2+3P)T_F/2 & while..N < n \le 2N \\ n((1+P)+(1+2P)+(1+3P))T_F/3 = n(1+2P)T_F/3 & while..2N < n \end{cases}$$

The value of n should be lower than 45. Assumed that N=15, P=0.3, TF=one single time unit. Then the result of simulation obtained by Matlab is shown in Figure 4.

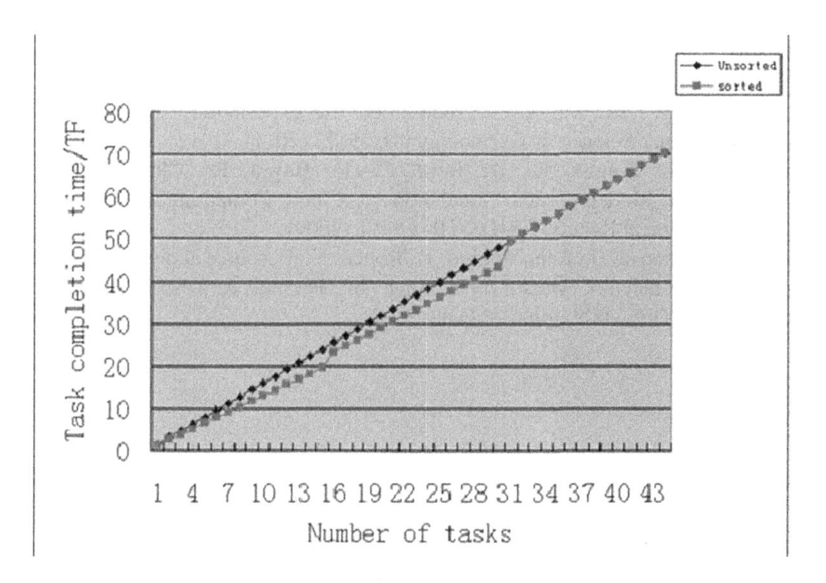

Fig. 4 The results of simulation

From figure 4 we can see that the scheduling algorithm based on Trust degree schedule the resources in order of their Trust degree. The risk time of completing tasks is significantly better than that no sorting while the number of tasks is not great.

5 Conclusion

With the development and application of Cloud Computing, the requirement of Cloud Computing service quality is increasing constantly. This paper research the resource scheduling strategy by introducing Trust degree. The analysis results show that the scheduling algorithm based on Trust degree is stable and reliable. The next goal is to find more effective resource scheduling algorithm in future.

Acknowledgment. This work is supported by the National Natural Science Foundation of China under Grant No. 71161007, the Social Science Fund Project of Ministry of Education under Grant No. 10YJCZH049, the Key Science and Technology Program of Haikou under Grant No. 2010-0067 and the Scientific Research Initiation Fund Project of Hainan University under Grant No. kyqd1042.

References

1. Song, Y.-J., Yang, X.-Z., Li, D.-Y., Cui, D.-H.: The Cloud Scheduler Politics of Multiprocessor Multitask Real Time Systems. Chinese Journal of Computers 23(10), 1107–1113 (2000)
2. Sun, R.-F., Zhao, Z.-W.: Resource Scheduling Strategy Based on Cloud Computing. Aeronautical Computing Technique 40(3), 103–105 (2010)
3. Liu, Z.-J., Zhang, T.-R., Xie, X.-C.: Cloud-based "user expectations" resource scheduling algorithm. Popular Science & Technology (4), 75–77 (2011)
4. Calheiros, R.N., Ranjan, R., De Rose, C.A.F., Buyya, R.: CloudSim."A Novel Framework for Modeling and Simulation of Cloud Computing Infrastructuresand Services." Technical Report, GRIDS-TR-2009-1 (2009)
5. Benoit, A., Casanova, H., Rehn-Sonigo, V., Robert, Y.: Resource allocation strategies for constructive in-network stream processing. In: Parallel & Distributed Processing Symposium (IPDPS 2009), pp. 1–8 (May 2009)

Sensor Data Filtering Algorithm for Efficient Local Positioning System

Sang-Seop Lim, Seok-Jun Ko, Jong-Hwan Lim, and Chul-Ung Kang*

Abstract. In this paper, we present a location recognition system for an entertainment robot and propose an idea to design a sensor network space. The sensor space consists of CDS (Cadmium sulfide) sensor cell of 24 by 24. Also our implemented hardware system is tested for setting a reference value. It is important to have exact location recognition. The algorithm for acquiring the reference value is proposed and its performance is evaluated with data of the implemented hardware system.

Keywords: Mobile platform, location recognition, sensor space, CDS (Cadmium sulfide) sensor.

1 Introduction

Unlike the past industrial robots, current robots require high technology such as human-friendly interface, interactive technology, voice recognition, object recognition, and user intention identifier. These technologies are called RT (Robot Technology) and regarded as one of the 21st century promising technologies along with IT, BT, and NT. Also, robots for personal, service, and welfare are now gaining more popularity. Especially, the robots for entertainment and educational purpose are actively being studied in recent years.

Sang-Seop Lim · Jong-Hwan Lim · Chul-Ung Kang
Department of Mechatronics Engineering
JEJU National University, Korea
e-mail: cukang@jejunu.ac.kr

Seok-Jun Ko
Department of Electronic Engineering
JEJU National University, Korea

* Corresponding author.

R. Lee (Ed.): Software Engineering Research, Management and Appl. 2012, SCI 430, pp. 185–194.
springerlink.com

These entertainment robots, compared with the other service robots, are mainly focused on performance such as dance and theater performances for leisure purpose. However, the movement range of entertainment robots is still limited and cannot be dynamic. In order to fulfill its task and behave properly, it is significantly important to have the capability to recognize their current location, environment, obstacles, and destination [1].

In order to have this location recognition, the following methods can be used; Gyro-sensors and Encoder are used to estimate the relative position [2]-[3], Ultrasonic and infrared sensors, vision sensor and GPS are used to point out the exact location [4], in recent years, mobile recognition robot using RFID system [7]-[8] that is based on the new concept called as "intelligent space" or "sensor network space"[5]-[6]. However, the above location recognition methods are required to redesign the system whenever the robot equipped various sensors.

In this paper, we propose the method to design the sensor network for performing an entertainment robot and present how to perform the location recognition without changing the entire robot structure using simple communication. The sensor space consists of CDS (Cadmium sulfide) sensor cell of 24 by 24. Also our implemented hardware system is tested for setting a reference value. It is important to have exact location recognition. The algorithm for acquiring the reference value is proposed and its performance is evaluated with data of the implemented hardware system.

The organization of this paper is as follows. In Section II, the sensor network space is introduced. The proposed algorithm for efficient local positioning system is explained in Section III. And Section IV shows experimental results of the system. Finally, the conclusion is presented in Section V.

2 Design the Sensor Space

In order to have location recognition, Fig. 1 shows the conventional system. Several sensors are equipped to estimate the relative position. This system requires much data processing due to many sensors, which is a critical problem in the

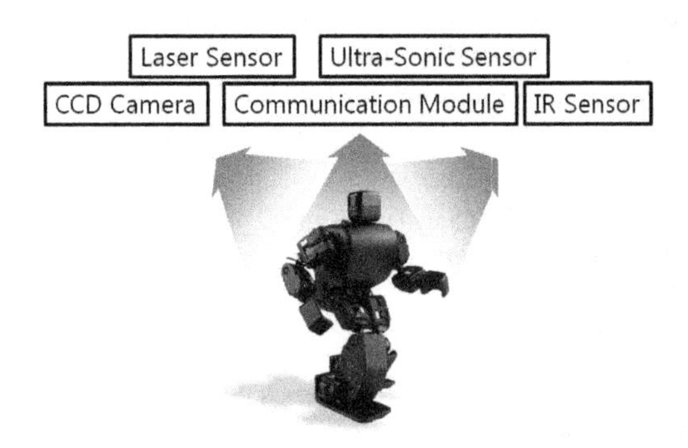

Fig. 1 The conventional location recognition system

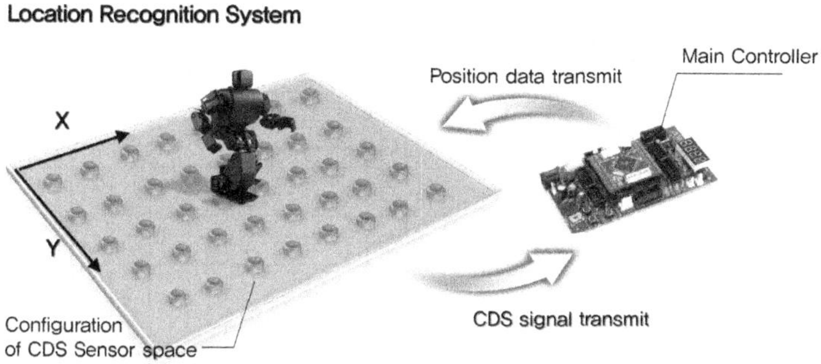

Fig. 2 The location recognition system by using Sensor network space

current method. If we want to utilize different type of mobile robots, we need to prepare different software for each robot according to the sensor. On the other hand, Fig.2 shows the proposed the location recognition system by using the sensor network space.

This system controls the robot position using CDS sensor data. The CDS sensor has the characteristic that the sensor value is changed depending on light changes. The process of this system is as follows;

1) The Sensor signals are transmitted to the main controller.
2) The main controller analyzes each local position of the robot and transmits the position data to the robot.
3) The robot is controlled depending on each local position.

As a result, this system can have location recognition without changing the entire robot structure using simple communication.

3 Proposed Algorithm for Efficient Local Positioning System

The sensor network system consists of the 24 by 24 matrix that is 576 CDS sensors. Same column of CDS sensor is connected to power lines and same row of CDS sensor is connected to signal line. Therefore, total power lines of 24(P1, P2 ... P23, P24) and signal lines of 24(S1, S2 ... S23, S24) are connected to Main controller. Fig. 3 shows the principle of the system. First, P1 power line is connected to the power. And Main controller can acquire signals of 24 about the P1 power line. And then Main controller switches the power line from P1 to P2. As a result, Main controller can acquire signals of 574 when the power line is sequentially switched from P1 to P24. We can convert this signal values to analog signal values in ADC of Main controller and can decide existence of the robot using analog signal values. However, it takes 864 ms to get analog signals of 576 CDS sensors due to the ADC's serial operation. Therefore, in order to decide the existence of the robot, it takes approximately 1 second. If the movement of the robot is very fast, the system cannot have location recognition properly. As a

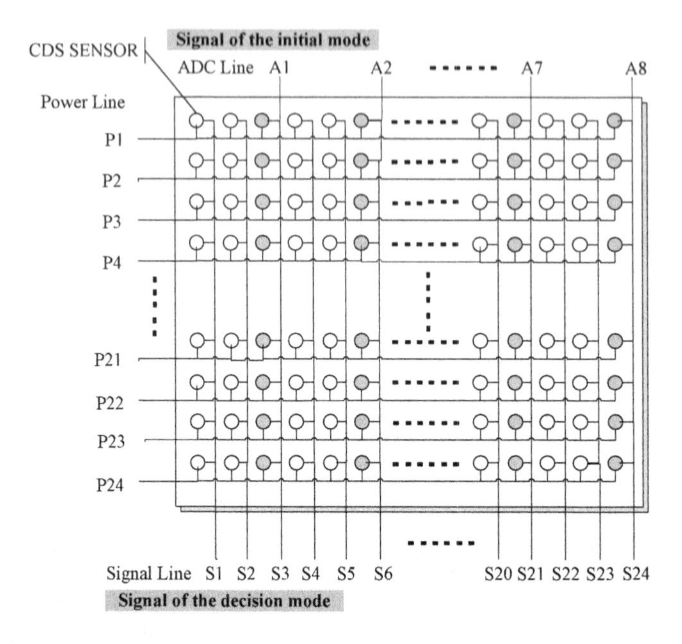

Fig. 3 The structure of the sensor network system

result, we need to convert analog signals to digital signals in that the system increases the speed. Fig.4 shows the method that converts analog signals to digital signals. The system has two modes that are the initial mode and the decision mode. ADC line of 8(A1, A2 … A7, A8) shown in Fig.3 is connect to Main controller in the initial mode. And then Main controller acquires analog signal. The role of the initial mode is the settings of the reference. And then the main controller compares the analog signals of the 576 CDS sensor with the set reference value. Finally, the main controller can decide the existence of the robot in the decision mode.

3.1 The Initial Mode

The main role of this initial mode as shown in Figure 4 is the setting the reference value of the comparator that is used in the decision mode. When the system is started, the main controller selects the initial mode, that is, the switch I is connected to the initial mode. And then, from the sensor matrix as shown in Figure 3, the main controller can acquire 8 sensor signals of each initial mode line (A1, A2 … A7, A8). However, every power line (P1, P2 … P23, P24) is not always operated. First, the P1 is on and others are not because the main controller supplies the power to the P1 only. Therefore the sensor signals of 8 on P1 column are converted to digital values by the ADC sequentially such as the followings; P1A1→P1A2→… →P1A8. The digital values are saved in a memory of the main controller. Second, the P2 is on and others are not. Similarly, the sensor signals of 8 on P2 column are converted to digital values (P2A1→P2A2→… →P2A8) and

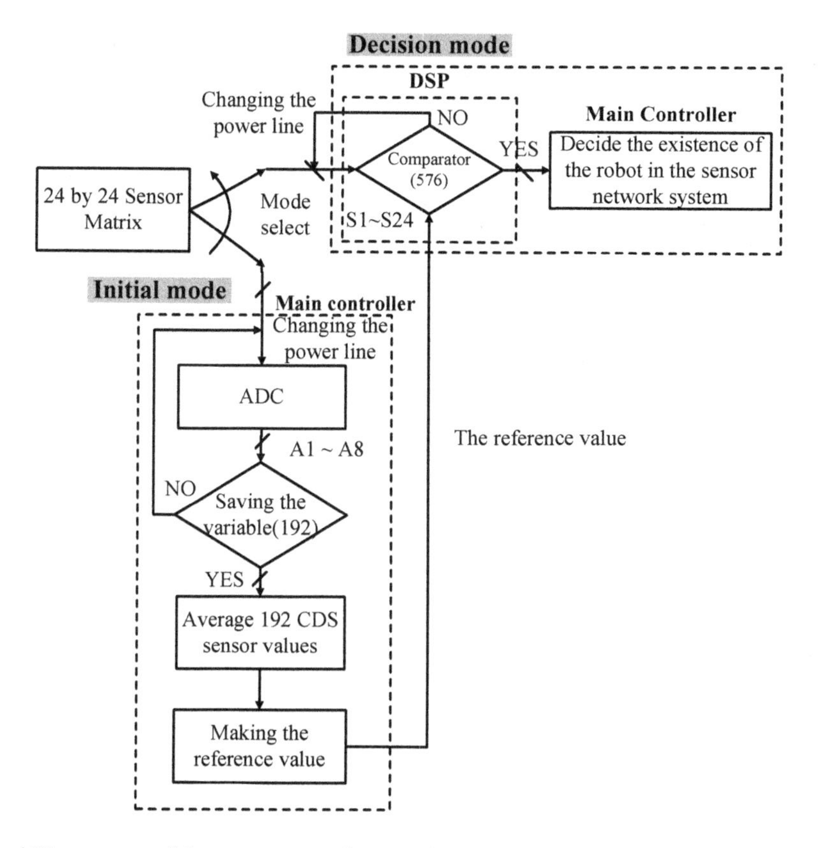

Fig. 4 The concept of the sensor network system's operation

saved. The main controller switches the power line from P1 to P24 sequentially. These processes are continued up to the P24. Finally, we can get an averaged value of 192 CDS sensor signals.

If the above process is repeated, as shown in Fig.5, we can get the instantaneous averaged values that are used in making the reference value. However, the values are varied due to the imperfect CDS sensor and unstable environment. Therefore we need some algorithm to make an optimal reference value.

The x-axis in Fig.5 indicates the second time. In our system, the consumption time for reading a sensor value is 1.5ms. It takes 288ms to make an averaged value because we can get the one averaged value by using 192 CDS sensor values. The result of the Fig.5 shows the averaged values as much as 300000.

In an unvarying environment, if the single averaged values are observed as [9].

$$x[i] = A + w[i], \quad i = 0,1,\ldots,N \tag{1}$$

where $w[i]$ is a Gaussian noise $N(0, \sigma^2)$ with zero mean and variance σ^2, then we can get the PDF (probability density function) as following

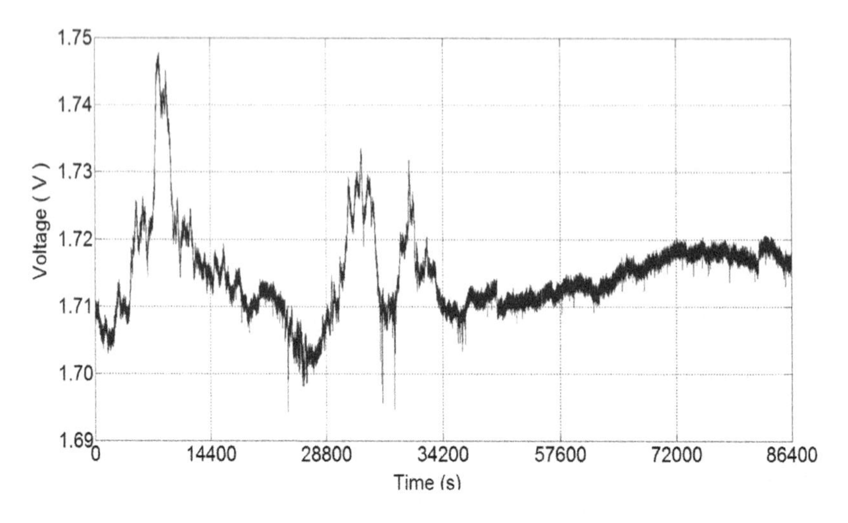

Fig. 5 The instantaneous averaged values of 192 CDS sensors

$$p(\mathbf{x};A)=\frac{1}{(2\pi\sigma^2)^{N/2}}\exp\left[-\frac{1}{2\sigma^2}\sum_{i=0}^{N-1}(x[i]-A)^2\right] \tag{2}$$

Taking the first derivative of the log pdf

$$\begin{aligned}\frac{\partial \ln p(\mathbf{x};A)}{\partial A}&=\frac{\partial}{\partial A}\left[-\ln[(2\pi\sigma^2)^{N/2}]--\frac{1}{2\sigma^2}\sum_{i=0}^{N-1}(x[i]-A)^2\right]\\&=\frac{1}{\sigma^2}\sum_{i=0}^{N-1}(x[i]-A)\\&=\frac{N}{\sigma^2}(\bar{x}-A)\end{aligned} \tag{3}$$

From the above equation, we can use a MVUE (Minimum Variance Unbiased Estimator) algorithm for making the reference value. Therefore, the MVUE in our system is

$$\bar{x}=\sum_{i=0}^{N-1}x[i] \tag{4}$$

The input value, $x[i]$, is the instantaneous averaged value in Voltage unit. The output value, $\bar{x}$, is the estimated reference value of A as shown in Fig. 6.

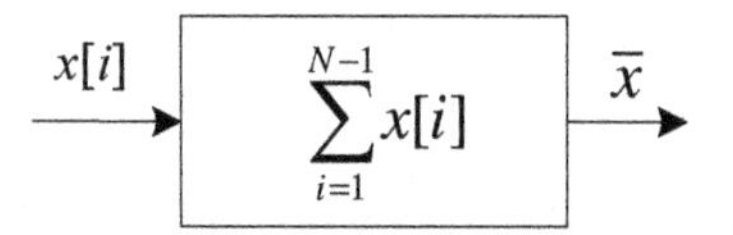

Fig. 6 The methods for acquiring the reference value

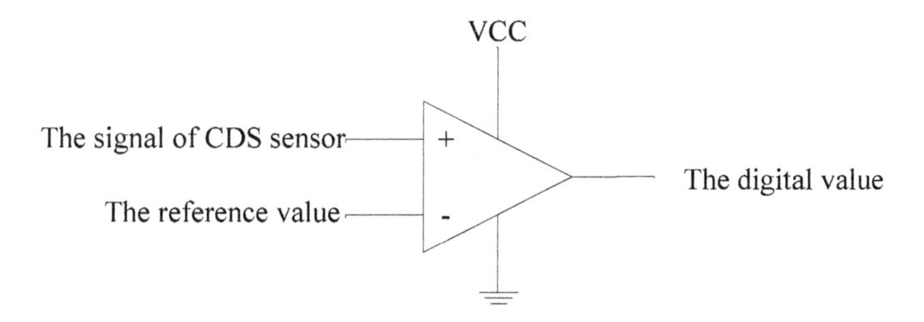

Fig. 7 The comparator

3.2 The Decision Mode

After the initial mode, the main controller selects the decision mode. The main controller selects all signal line (S1~S24) unlike the initial mode. Each signal line is connected to comparator as shown in Fig. 7. The main controller sequentially switches the power line from P1 to P24 and every acquired analog value, depending on the power line, is compared with the reference value. If the sensor signal is below the reference value, output is 1 according to the principle of the comparator. Therefore, we can decide the existence of the robot depending on output value.

4 Experimental Results

The Experimental Environment is the office room of approximately 350 lux as shown in the Fig. 8. The office room keeps out of disturbance. In its current condition, the sensor filtering algorithm performed the 100 cycles.

Fig.9 shows the Gaussian distribution. When the system performs 30 cycles, Fig. 9 shows the best distribution between 1.72v to 1.73v. Also, we can find the averaged value 1.727v. Therefore, we can set up the reference value in 50% of 1.727v.

Fig.10 and 11 shows the graph of Mean and Variance after using Algorithm of 30 cycles. Between 0v and 5v, mean values approach to 1.722v and Variance values approach to minimum value in Algorithm of 20 cycles. If the condition of the experimental environment is not changed rapidly, mean value and variance value is constant after 20 cycles. As a result, in order to acquire the optimal reference value, the system will perform the minimum of 20 cycles in the initial mode.

It takes 1.5ms to acquire signal of one CDS sensor, and 288ms to acquire analog data of 192, and 5760ms to acquire analog data of 20 cycles. Therefore, if the system is started, first, Main controller can acquire the reference value in the initial mode at 5.7 second, and then Main controller can decide the existence of the robot in decision mode.

Fig. 8 The Experimental Environment

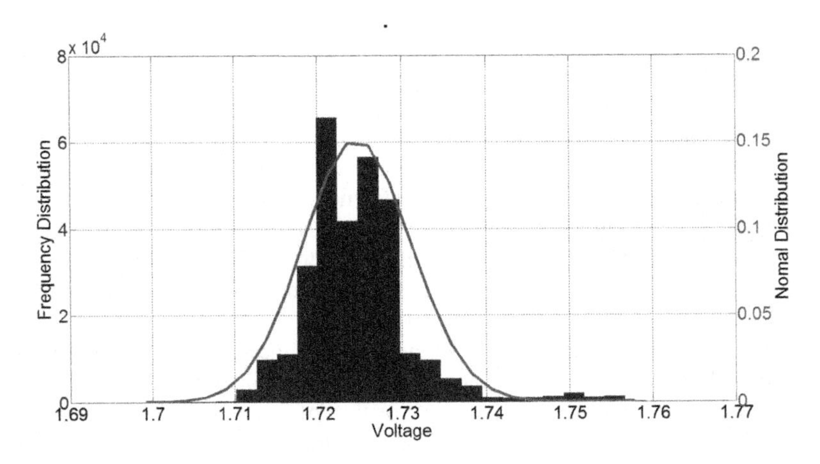

Fig. 9 The Gaussian pdf from the experimental data

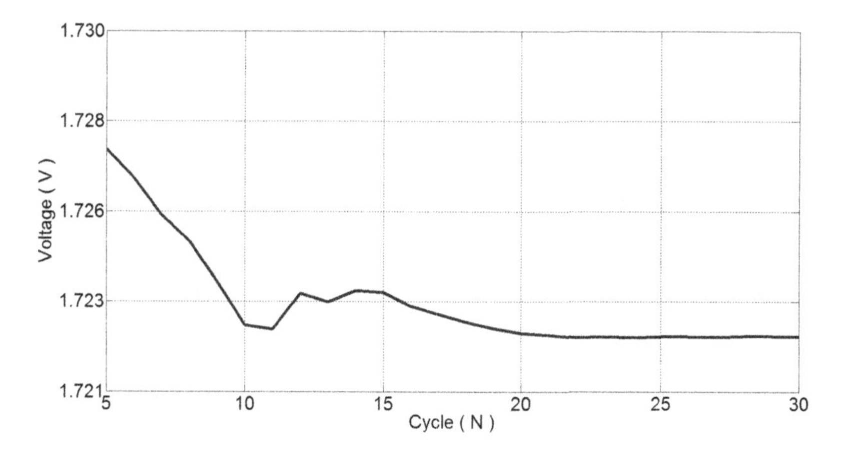

Fig. 10 Mean values depending on N cycle

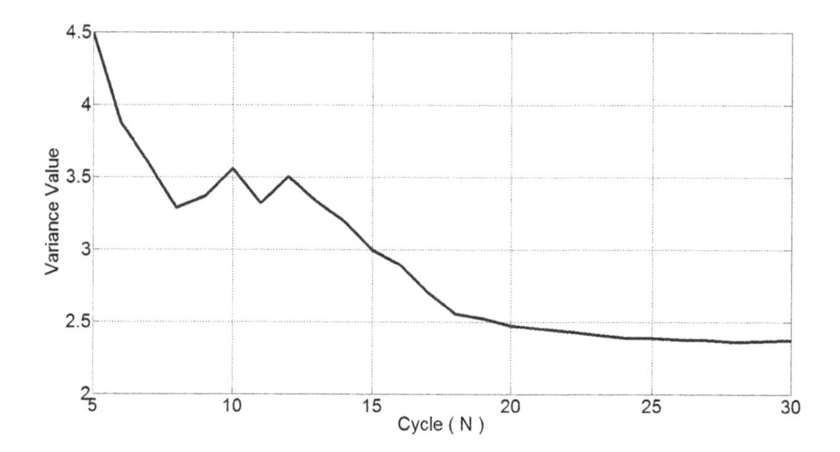

Fig. 11 Variance value depending on N cycle

5 Conclusion

In this paper, we suggest the method to design the sensor space for performing an entertainment robot. And, we present how to perform the location recognition with the sensor network space unlike the conventional system. The sensor space consists of CDS sensor cell of 24 by 24. In order to have exact location recognition, we propose the algorithm for acquiring the reference value.

Also our implemented hardware system is tested for setting a reference value. It is important to have exact location recognition. The algorithm for acquiring the reference value is proposed and its performance is evaluated with data of the implemented hardware system.

However, the reference value of system must to be updated frequently because the CDS sensor is sensitive to the light change and has different characteristics in each sensor. In further work, it is expected that the Robot has the exact location recognition by using the new algorithm with good performance.

References

1. Lee, S.: Using Vision Sensor and Sonar of Mobile Robot Localization. Chunguam National University (2001)
2. Komoriya, K., Oyama, E.: Position estimation of a mobile robot using optical fiber gyroscope. In: Proc., IEEE/RSJ/GI International Conference on Intelligent Robots and Systems, IROS 1994, September 12-16, vol. 1, pp. 143–149 (1994)
3. Lee, S., Song, J.-B.: Robust mobile robot localization using optical flow sensors and encoders. In: Proc. IEEE International Conference on Robotics and Automation, ICRA 2004, vol. 1(26), pp. 1039–1044 (April 2004)
4. Ro, Y.S., Park, K.S.: A study of Optimal Position Estimation for Mobile Robots. Journal of Engineering Research 26(1), 243–257 (1995)
5. Lee, J.-H., Hashimoto, M.: Controlling Mobile Robots in Distributed Intelligent Sensor Network. IEEE Transaction on Industrial Electronics 50(5), 890–902 (2003)
6. Morioka, K., Lee, J.-H., Hashimoto, H.: Human-Following Mobile Robot in a Distributed intelligent Sensor Network. IEEE Transaction on Industrial Electronics 51(1), 229–237 (2003J)
7. Yamano, K., Tanaka, K., Hirayama, M., Kondo, E., Kimuro, Y., Matsumoto, M.: Self-Localization of Mobile Robots with RFID System by using Support Vector Machine. In: Proc. IEEE/RSJ Int. Conf. Intelligent Robots, Systems, pp. 3756–3761 (2004)
8. Li, L.M., Liu, Y., Lau, Y.C., Patil, A.P.: LANDMARC: indoor location sensing using active RFID. In: Proc. IEEE Int. Conf. Pervasive Computing and Communications, March 26-26, pp. 407–415 (2003)
9. Kay, S.M.: Fundamentals of Statistical Signal Processing; Estimation Theory. Prentice Hall (1993)

Author Index